Contaminant Transport
A Numerical Approach

by D. James Benton

Copyright © 2021 by D. James Benton, all rights reserved.

Forward

In previous texts we have covered ***Computational Fluid Dynamics***, ***Mass Transfer***, ***Plumes***, and ***Particle Tracking***. Contaminant Transport is a specific application, involving diffusion, dispersion, and advection. While movement of the transporting media (e.g. groundwater, surface water, or air) is important, this text will not cover those details in depth. The reader should be familiar with fluid flow and how it may be calculated before delving into the current subject mater. We cover passive as well as decaying substances. While this text does cover theory, it is primarily a compilation of examples based on actual remediation projects. Accurate modeling is essential for containment or capture, for we must know where the contaminant is going and when it will get there if we are ever to achieve effective remediation.

All of the examples contained in this book,
(as well as a lot of free programs) are available at...
https://www.dudleybenton.altervista.org/software/index.html

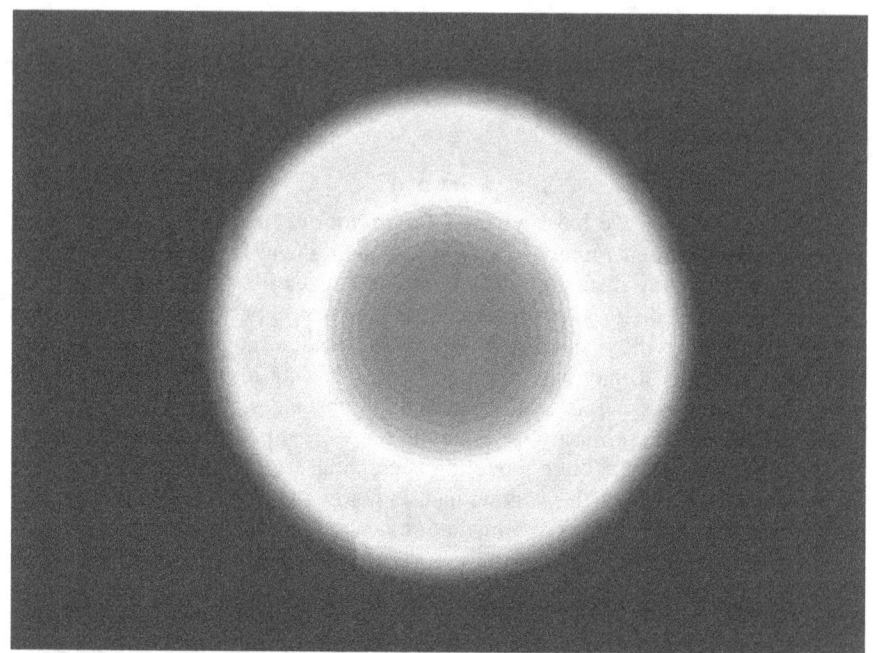

Figure 1. Analytical Solution

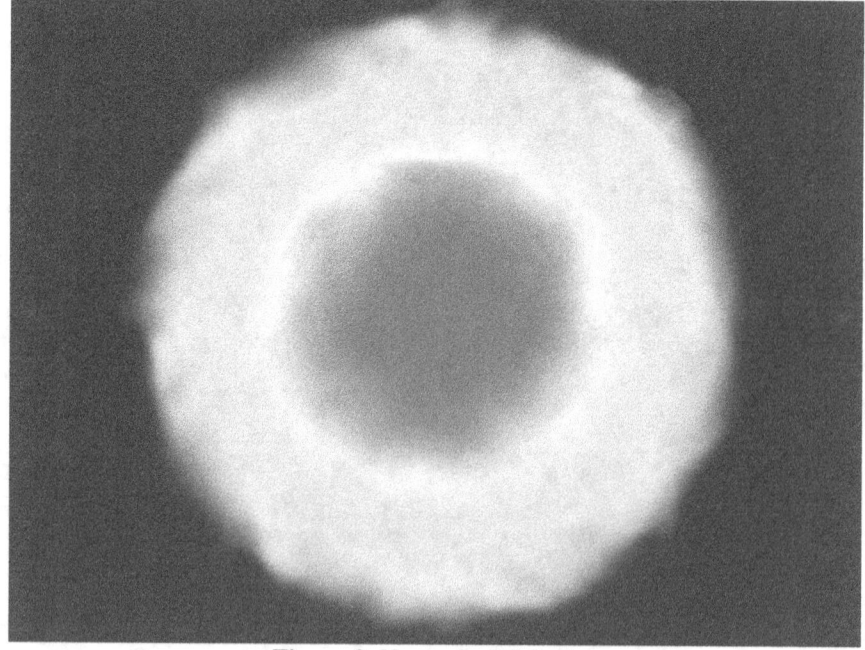

Figure 2. Numerical Solution

Table of Contents

	page
Forward	i
Chapter 1. Diffusion in 2D	1
Chapter 2. Dispersion in 2D	11
Chapter 3. Advection in 2D	15
Chapter 4. Contaminant Transport in 3D	27
Chapter 5. MODFLOW Based Models	35
Chapter 6. FRAC3D Based Models	55
Chapter 7. Pump-and-Treat	79
Chapter 8. Reaction and Decay	83
Chapter 9. AT123D Analytical Solution	87
Appendix A. Displaying Data in 3^+D	93
Appendix B. 3D Data files for Tecplot™	97
Appendix C: 3D Data Files for TP2	99
Appendix D. Build3D Model Builder	101
Appendix E. Initial Concentrations	105

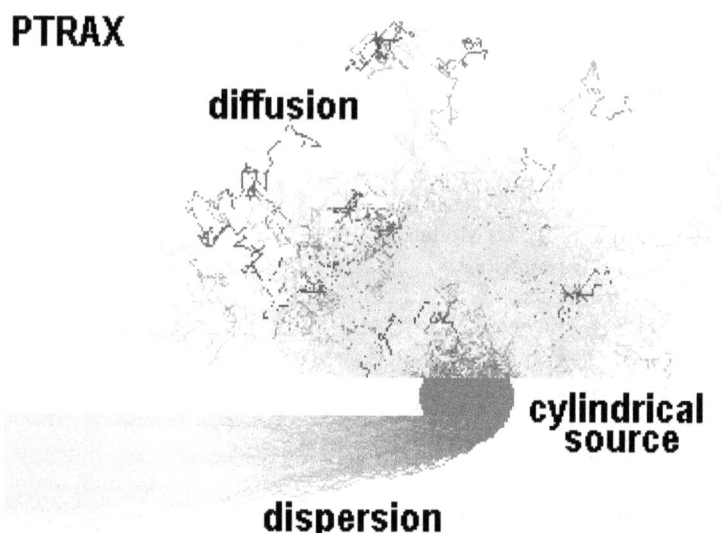

Figure 3. Particle Tracks Showing Diffusion and Dispersion

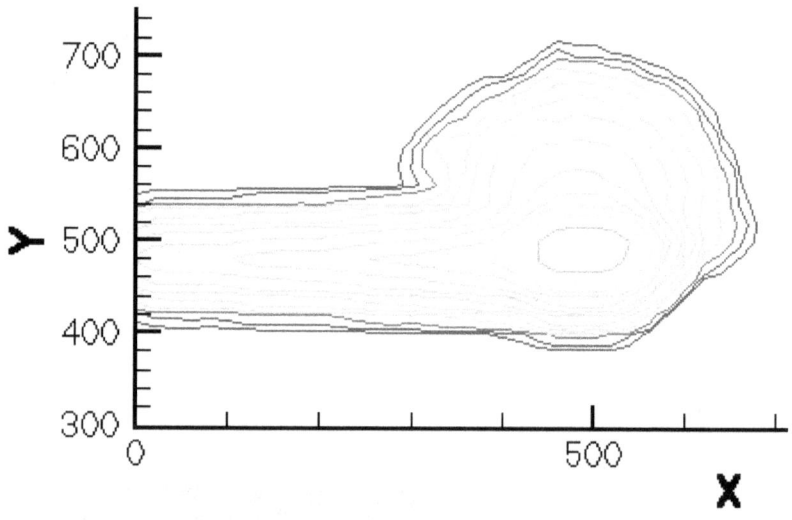

Figure 4. Corresponding Concentration Contours

Chapter 1. Diffusion in 2D

We begin with what is perhaps the simplest meaningful example: diffusion in two dimensions. We have already covered one-dimensional diffusion in *Mass Transfer*. If the diffusion coefficient, D, is constant (spatially and temporally), Fick's second law becomes:

$$\frac{\partial C}{\partial t} = D\left(\frac{\partial^2 C}{\partial x^2} + \frac{\partial^2 C}{\partial y^2}\right) \qquad (1.1)$$

The simplest solution to implement is finite differences spatially and Euler's (explicit, single-step) method temporally. Using subscripts i and j to indicate the X and Y directions, respectively, Equation 1.1 becomes:

$$\frac{C_{i,j}^{t+\Delta t} - C_{i,j}^t}{\Delta t} = D\left(\frac{C_{i+1,j}^t - 2C_{i,j}^t + C_{i-1,j}^t}{\Delta x^2} + \frac{C_{i,j+1}^t - 2C_{i,j}^t + C_{i,j-1}^t}{\Delta y^2}\right) \qquad (1.2)$$

If the problem is not too stiff (rapidly changing spatially or temporally resulting in numerical instability) and we use a sufficiently fine grid and small time steps, we can simply march forth in time to obtain a solution. We must choose a contaminant substance, properties, scale (length and time), and initial concentration. I have provided modeling for the remediation of several sites where Trichloroethylene (TCE) is involved and so we select this for our first example. TCE is a contaminant of particular interest to the USEPA and many articles related to this common solvent can be found on their website.

The solubility of TCE in water is approximately 1000 mg/l[1] and the density of TCE is approximately 1.5 times that of water so an initial concentration of 300 ppm is reasonable. We will begin with an amorphous media before considering porosity and the separate phases (solid and liquid). The diffusion coefficient for TCE in water is approximately 0.00001 cm²/sec. The initially contaminated area extends over 80m and is triangular in shape. The domain of consideration is 640m by 640m. The grid spacing in both directions is 1m.

The ratio $D/\Delta x^2 \approx 0.3$/yr provides an estimate of the appropriate time step. The dimensions (640x640) facilitate graphical representation of the results, which will be two-dimensional, sequential time frames, colored from blue to red based on the log of the concentration with red indicating log(300), blue indicating log(0.3), and gray indicating zero. All of the files can be found in the

---

[1] "Using the Combined SESOIL/AT123D Models to Develop Site-Specific Impact to Ground Water Soil Remediation Standards for Mobile Contaminants," Guidance Document Version 2.1, New Jersey Department of Environmental Protection, Trenton, New Jersey, May, 2014. [As well as providing various properties, this is an important reference for using AT123D, which we discuss in Chapter 9.]

online archive in folder examples\TCE2D. The initial conditions are shown in the following figure:

Figure 5. Example 1 Initial Conditions

Using the simple forward Euler method for the time step (Equation 1.2) it should be clear that the contaminant cannot spread any more rapidly than one nodal point per time step. Considering this fact and also the ratio of the diffusion coefficient to the grid size, the time step should be no more than $\Delta t<0.3$ yr. We initialize the field and then march forward in time. While we could immediately update each concentration while running through a nest of two loops (X and Y), this would produce a sweeping artifact as the cells are updated. Instead, we allocate a working array to contain $\partial C/\partial t$ and then implement $C=C+(\partial C/\partial t)\Delta t$.

It is not necessary to save an image at each time step. One frame per 100 time steps is often enough so that the bitmap image need only be updated every 30 years. As we are coloring the image with the log of the concentration, once the contaminant spreads, it may take longer to update the image than to advance the solution. After 1800 years the contaminant has spread as shown in the following figure. Note that the concentrations are logarithmic.

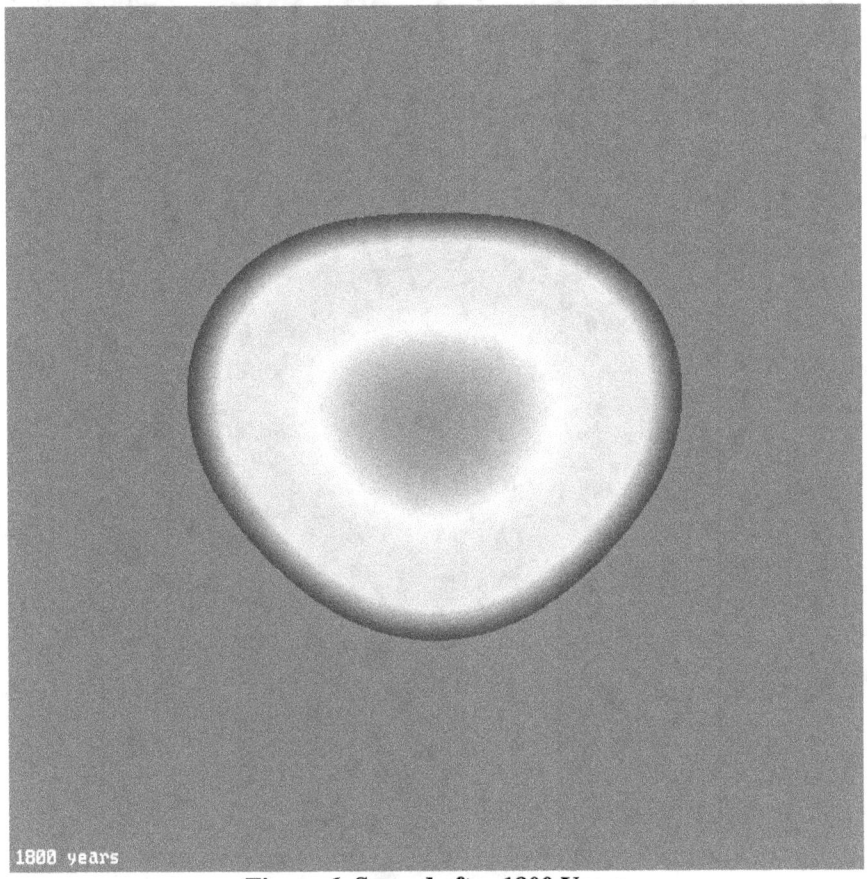

Figure 6. Spread after 1800 Years

You may wonder how small of a time step is small enough and is the forward Euler method adequate. If we increase the time step from 0.3 years (recall $D/\triangle x^2 \approx 0.3/yr$) to 1.0 yr, the solution quickly becomes unstable. If we used a higher order method (for example, 4th Order Runge-Kutta, rather than forward Euler), we could use a somewhat larger $\triangle t$, but not much. We would also have the problem of the contaminant not spreading more than one node (in this case 1m) in any direction per time step. This is not to say that higher order methods are without advantage, only that for this simple problem, they do not offer a significant benefit. These will be used in subsequent examples. Pure diffusion is inherently stable. When we introduce flow and turbulence, we will be more concerned with stability and will resort to more clever methods, which will also be more computationally intensive. The unstable result after 300 years with a time step of 1 yr is shown in this next figure:

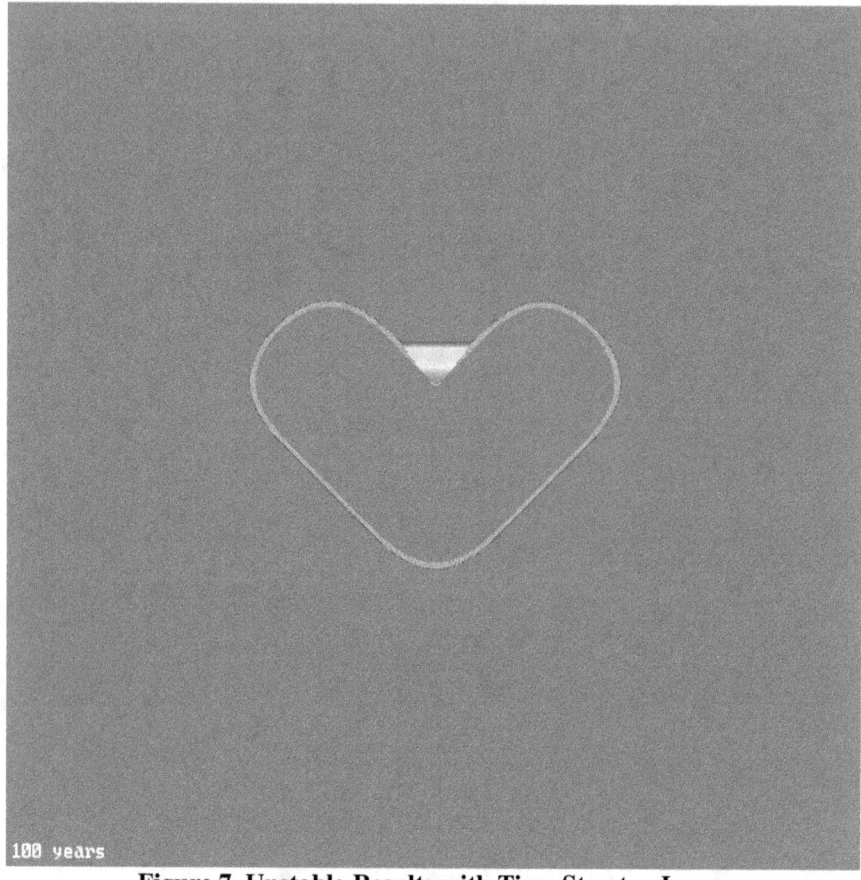

Figure 7. Unstable Results with Time Step too Large

Boundaries

Equation 1.2 can only be applied at the interior points. The code is quite simple. Recall the working array, W, used to calculate the change in C, as listed below:

```
void AdvanceSolution(double dt)
   {
   int i,x,y;
   for(y=1;y<Ny-1;y++)             /* interior points */
      {
      for(x=1;x<Nx-1;x++)
         {
         i=Nx*y+x;
```

```
        dCdt[Nx*y+x]=a*((C[i+ 1]-2.*C[i]+C[i- 1])
                +(C[i+Nx]-2.*C[i]+C[i-Nx]));
        }
    }
    for(y=1;y<Ny-1;y++)              /* interior points */
    {
        for(x=1;x<Nx-1;x++)
        {
            i=Nx*y+x;
            C[i]+=dt*dCdt[i];
        }
    }
```

This still leaves all of the nodes around the edges (top, bottom, left, right, plus four corners). In this case "natural" boundary conditions are applied, which means $\partial C/\partial x=0$ on the left and right sides and $\partial C/\partial y=0$ on the top and bottom sides. The most "natural" way of handling the corners is to set these nodes equal to the average of the three closest interior nodes. Details can be found in the code (tce2d.c).

```
for(y=0,x=1;x<Nx-1;x++)              /* bottom side */
    C[Nx*y+x]=C[Nx*(y+1)+x];
for(y=Ny-1,x=1;x<Nx-1;x++)           /* top side */
    C[Nx*y+x]=C[Nx*(y-1)+x];
for(x=0,y=1;y<Ny-1;y++)              /* left side */
    C[Nx*y+x]=C[Nx*y+x+1];
for(x=Nx-1,y=1;y<Ny-1;y++)           /* right side */
    C[Nx*y+x]=C[Nx*y+x+1];
x=y=0;                               /* bottom left corner */
C[Nx*y+x]=(C[Nx*y+x+1]+C[Nx*(y+1)+x]+C[Nx*(y+1)+x+1])/3;
x=Nx-1;y=0;                          /* bottom right corner */
C[Nx*y+x]=(C[Nx*y+x-1]+C[Nx*(y+1)+x]
        +C[Nx*(y+1)+x-1])/3.;
x=0;y=Ny-1;                          /* top left corner */
C[Nx*y+x]=(C[Nx*y+x+1]+C[Nx*(y-1)+x]
        +C[Nx*(y-1)+x+1])/3.;
x=Nx-1;y=Ny-1;                       /* top right corner */
    C[Nx*y+x]=(C[Nx*y+x-1]+C[Nx*(y-1)+x]+C[Nx*(y-1)+x-1])/3.;
```

Impact of Properties

We next consider what happens when we double the diffusion coefficient. For illustration, we will double it in the Y direction but not the X. We must half the time step because we must consider the most restrictive case (X or Y). After the same 1800 years the results are shown in Figure 4, along with the outline of the extent of contamination from Figure 2. Again, recall that the colors represent the log of the concentration. One question that often arises in modeling is, "What if we're not sure about the properties (e.g., diffusion coefficient)?" We may not be sure (especially underground) or precise (properties may vary and "undisturbed" soil samples are hard to obtain), but this doesn't necessarily translate into the same level of uncertainty in the results.

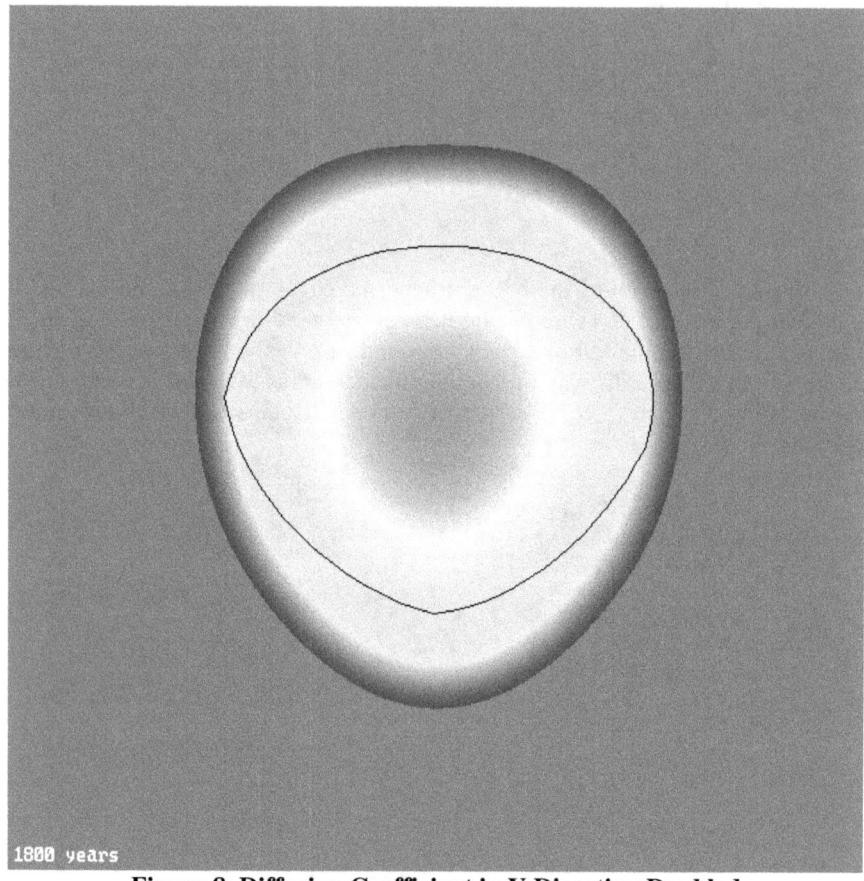

Figure 8. Diffusion Coefficient in Y Direction Doubled

Properties, such as diffusion coefficient, describe the behavior of the media, which is often underground. Soil properties rarely vary linearly. More often these vary with geological formation or alluvial deposition. While these often vary vertically (e.g., layers of different soil types), the planar (X and Y for this example) variation is separated into regions. The model builder, Build3D, which we will discuss in a later chapter, accepts soil types in various geometric shapes and builds the properties so indicated into the corresponding grid elements. In this example, we will just use two polygons and test for above or below and right or left, so as to assign a high, medium, and low value of diffusion coefficient, as shown in the next figure.

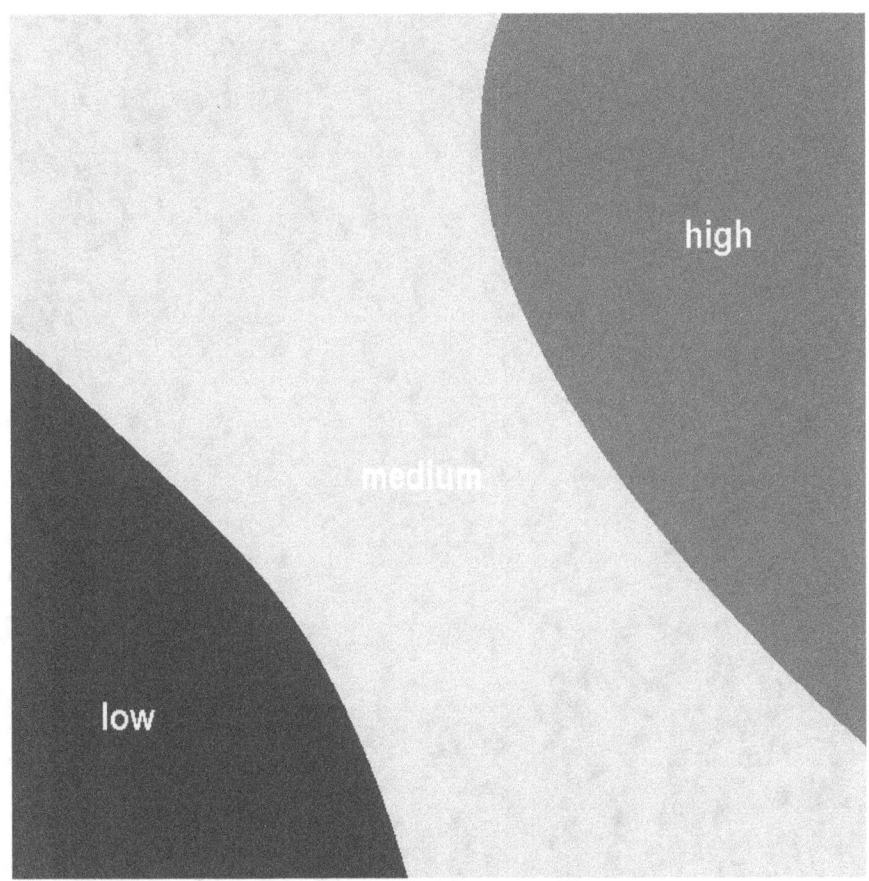

Figure 9. Typical Property Zones

We expect the contaminant will preferentially spread up and to the left into the high area, continue as before in the mid zone and less penetration into the low zone in the lower left. This is exactly what we see with real contaminant plumes. In fact, rate of spread of tracer chemicals is often used to infer properties of the soil. One of the longest studies conducted on this subject is the Macro Dispersion Experiment (MADE) conducted at Columbus Air Force Base. I served as the applied mathematician on the project throughout the 1990s, writing much of the software used. Many reports can be found on this experiment, published by EPRI and also the USGS. I continued working with some members of the same Team for several years after that, using the skills, knowledge, and tools developed during MADE on other projects for the USGS, USEPA, USDoE, USDoD, and USACoE. The resulting contaminated area after 1800 years is shown in this next figure.

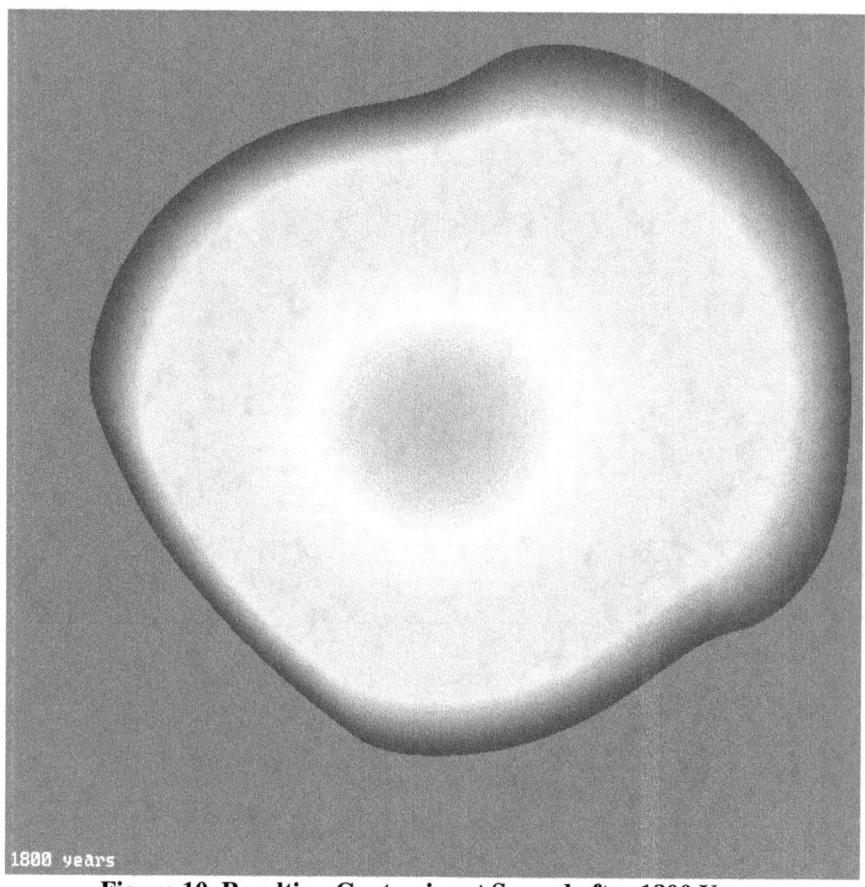

Figure 10. Resulting Contaminant Spread after 1800 Years

<u>Time Steps and Stability</u>

For this simulation (using the diffusion field shown in Figure 5 and resulting in the spread after 1800 years shown in Figure 6), we used a time step of Δt of 0.05 (1/20th) year. The highest diffusion coefficient in Figure 5 is 0.0006 cm²/sec. This combined with the grid spacing of 1m yields a characteristic ratio of approximately 1/10th ($\Delta t D/\Delta x^2 = 0.095$). If we double the diffusion coefficient throughout the domain, we should expect to also divide the time step in half to 0.025 (1/40th) year in order to maintain stability. Not surprisingly, the resulting spread is roughly twice that shown in Figure 6.

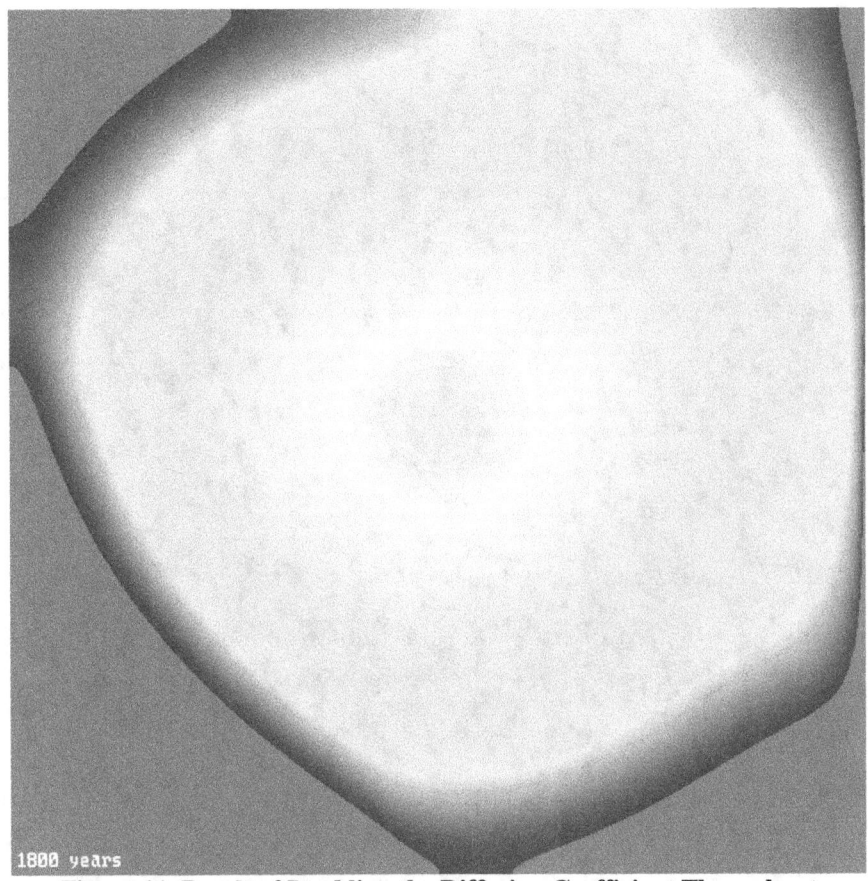

Figure 11. Result of Doubling the Diffusion Coefficient Throughout

If we had not reduced the time step in proportion to the maximum diffusion coefficient, the solution would have eventually become unstable. What happens numerically for this partial differential equation (i.e., Laplace's) is an ever-increasing oscillation (high/low up/down) of values (in this case concentrations) at adjacent nodes, which is a numerical artifact and completely uncharacteristic of the underlying physics described by the governing equation. In other words: garbage. This isn't always easy to detect when running a model. Because this particular model generates images as it steps along through time, we can see this "garbage" show up. A mere factor of 2 (in the wrong way, larger time step or larger diffusion coefficient) will quickly veer of course in 60 years trashing most of the domain, as shown in this next figure:

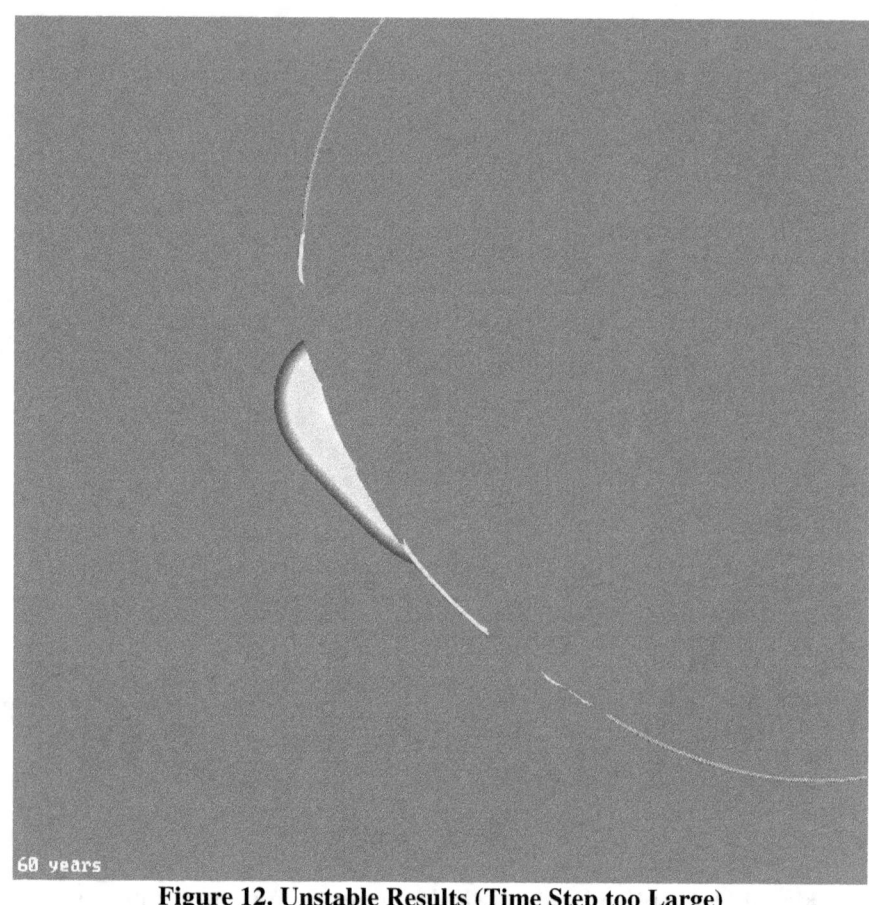
Figure 12. Unstable Results (Time Step too Large)

Chapter 2. Dispersion in 2D

If contaminants only spread by diffusion and the properties used for the examples in Chapter 1 are at all reasonable, contamination in groundwater would be of much lesser concern. In real situations, the contaminants move much faster than this first example would indicate because there are more mechanisms available. While diffusion is mostly omnidirectional (except for markedly different soil types), dispersion exhibits a preferential direction. Increased dispersive spreading (compared to diffusive spreading) perpendicular to the preferential direction is often seen, though noticeably less than that along the preferential direction. The same partial differential equation governs both diffusion and dispersion and the respective coefficients have the same units. A simple variation in dispersion coefficient (which we assign the symbol, A) along the X-axis is shown below (increasing toward the bottom right corner):

Figure 13. Dispersion Coefficient in the X Direction

A different (though still simple) variation in dispersion coefficient in the Y direction (with the same linear coloring scale) is shown in this next figure (decreasing toward the bottom right corner):

Figure 14. Dispersion Coefficient in the Y Direction

These dispersion coefficients are approximately 10x that of the diffusion coefficients (a reasonable ratio), implying that the rate of spreading due to dispersion will be about 10x that of diffusion. The dispersion coefficient is often expressed as the product of the dispersivity (having units of length) times the pore water velocity (having units of length/time) to obtain the dispersion coefficient (having units of length²/time). There will also be more spatial asymmetry arising from the combination of non-uniform properties assigned over the domain. The time step must be adjusted accordingly in order to keep the solution from becoming unstable. We have added two steps in the code (tce2d.c) to account for this automatically: 1) keeping track of the maximum value of A or D, and 2) calculating the value of Δt to make the ratio equal to one-third ($\Delta t D/\Delta x^2 = 1/3$). We also calculate the frequency of saving results so that the frames are written at convenient values of time. After only 180 years under this new scenario, the concentration field becomes:

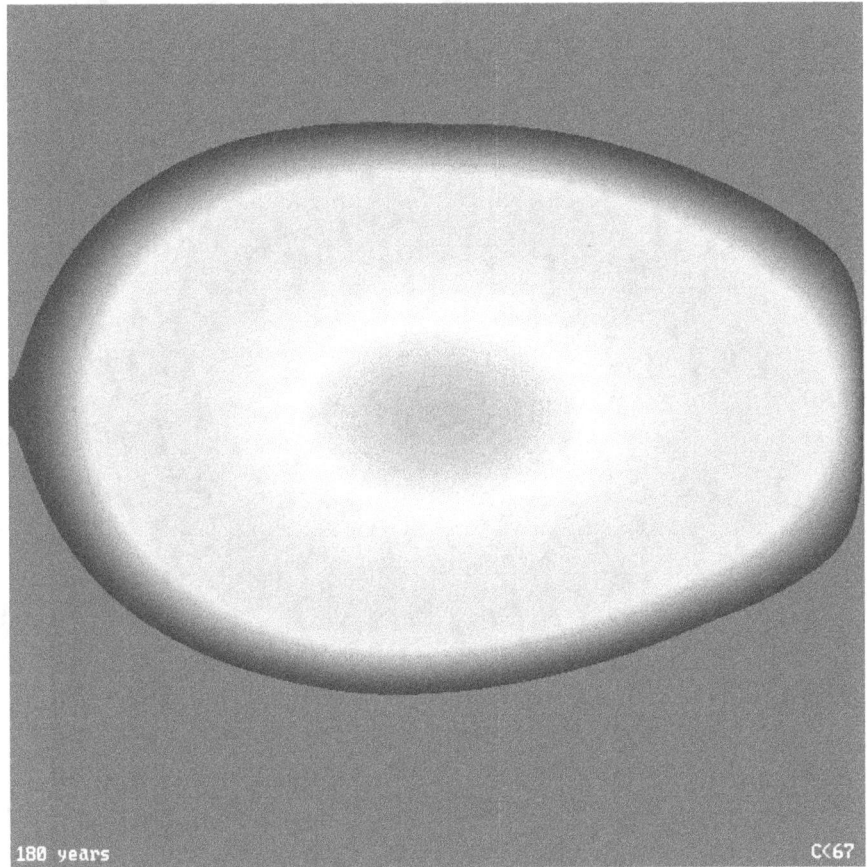

Figure 15. Concentrations after 180 Years with Diffusion and Dispersion

By adding approximately 10x dispersion to diffusion, in one-tenth the time (180 vs. 1800 years), the contaminant has spread so much that the red level has completely disappeared from the figure. The maximum concentration has fallen from 300 ppm to 67 ppm and even that level extends over a smaller area. Keep in mind that mass is conserved (we have not yet considered reactions or decay) so that the same amount of contaminant is still in the environment, only it is much more spread out. If there were any hope of containing or extracting it, this scenario would be a complete failure. Accurate modeling is essential for containment or capture. We must know where the contaminant is going and when it will get there if we are ever to remediate this situation.

Chapter 3. Advection in 2D

We now consider advection; that is, movement of the fluid media containing the contaminant. Before we do this, we must consider porosity, which we ignored in the first two chapters. When considering a contaminant in the ground, there are at least three materials: the contaminant, the soil, and groundwater. We must, therefore, consider spreading (and possibly reaction) of the contaminant in the water and the soil, which may be quite different. We first consider only the contaminant and water, ignoring the soil.

Porosity

Porosity is given the symbol, φ, and is the volume of the secondary material (in this case water) over the total volume. Porosity is the fraction of the total space between the pieces of gravel in the figure below. Sand is much smaller and packs more closely than gravel so that the porosity is much less. Silt and clay are even finer, pack even more closely, and have proportionately lower porosities. In this chapter we will consider the water as passing through the spaces between the soil particles, which do not participate in the processes except to channel and impede the flow of water. We sample the water by pumping it out of wells and analyzing it for the presence of contaminants. These concentrations provided by a laboratory are in the water by weight or volume, not in the total volume (water plus soil), a difference we will discuss later.

Figure 16. Typical Gravel Illustrating Porosity

Porosity can range from 0 to 1, but most often is between 0.2 and 0.3. Typical porosities for some common soil types are listed in Table 1.

Table 1. Typical Soil Porosities[2]

USDA Soil Class	porosity
clay (very fine)	0.20
clay (fine)	0.22
sandy clay	0.24
sandy loam	0.25
silty clay	0.25
sandy clay loam	0.26
silt	0.27
silty clay loam	0.27
loamy sand	0.28
clay loam	0.30
loam	0.30
sand	0.30
sily loam	0.35

Governing Equation Revisited

We now consider the governing partial differential equation (which is the conservation of mass of the contaminant for a differential control volume) in more detail. It is helpful to introduce here the del operator, ∇, to abbreviate the spatial derivatives while implying all three dimensions, which will facilitate our eventual advancement into 3D. This useful operator is defined as:

$$\nabla = \hat{i}\frac{\partial}{\partial x} + \hat{j}\frac{\partial}{\partial y} + \hat{k}\frac{\partial}{\partial z} \qquad (2.1)$$

In Equation 2.1, $\hat{i}$, $\hat{j}$, and $\hat{k}$ are unit vectors in the X, Y, and Z directions, respectively. When applied to a scalar quantity, such as concentration, the del operator yields a vector result. The diffusion and dispersion contributions (to the conservation of mass of the contaminant for a differential control volume) should more accurately be written:

$$\nabla \bullet [(A+D)(\nabla \varphi C)] \qquad (2.2)$$

On the right side of Equation 2.2, the terms A, D, φ, and C are all scalar quantities, but the del operator (in the middle of this group) makes the result a vector. The symbol ($\bullet$) indicates the dot product. The dot product of two vectors is a scalar so that the entire expression is a scalar. The properties (dispersion

---

[2] Bonazountas, M. and Wagner, J. M., "SESOIL: A Seasonal Soil Compartment Model," USEPA Report PB86112406, 1984.

coefficient, diffusion coefficient, and porosity) are inside the one or more differential operators (one for A and D; two for φ). When we introduced a spatially varying diffusion coefficient in Chapter 1, we should have written the term in the expanded form of Equation 2.3 and also modified the finite difference equations in the code accordingly.

$$\frac{\partial}{\partial x}\left(D\frac{\partial C}{\partial x}\right)+\frac{\partial}{\partial y}\left(D\frac{\partial C}{\partial y}\right) \quad (2.3)$$

Adding the dispersion coefficient and also the porosity, this term expands to:

$$\frac{\partial}{\partial x}\left[(A+D)\frac{\partial(\varphi C)}{\partial x}\right]+\frac{\partial}{\partial y}\left[(A+D)\frac{\partial(\varphi C)}{\partial y}\right] \quad (2.4)$$

The advective contribution to the conservation of mass of the contaminant in vector form using the del operator is given by:

$$\nabla \bullet (\varphi C \vec{V}) \quad (2.5)$$

Again, the right side of Equation 2.5 is a vector, as porosity, φ, and concentration, C, are scalars and velocity, $\vec{V}$, is a vector. As before, the dot product yields a scalar, which becomes:

$$\frac{\partial}{\partial x}(\varphi C u)+\frac{\partial}{\partial y}(\varphi C v) \quad (2.6)$$

This expands to:

$$u\frac{\partial}{\partial x}(\varphi C)+v\frac{\partial}{\partial y}(\varphi C)+\varphi C\left(\frac{\partial u}{\partial x}+\frac{\partial v}{\partial y}\right) \quad (2.7)$$

The last term in Equation 2.7 is zero by virtue of continuity. Combining these terms (diffusion plus dispersion and advection), we obtain an expression for the conservation of mass for the contaminant:

$$\frac{\partial(\varphi C)}{\partial t}=\frac{\partial}{\partial x}\left[(A+D)\frac{\partial(\varphi C)}{\partial x}\right]+\frac{\partial}{\partial y}\left[(A+D)\frac{\partial(\varphi C)}{\partial y}\right]-\left[u\frac{\partial}{\partial x}(\varphi C)+v\frac{\partial}{\partial y}(\varphi C)\right] \quad (2.8)$$

We will expand and adapt the previous example (tce2d.c) to account for the expanded differentials and also the additional terms. Instead of further complicating this code and obscuring the earlier features, we will slightly modify the name. Tetrachloroethylene (PCE) is often found with or instead of

TCE and so this example will be called pce2d.c and retain all the other properties, as these are quite similar for the two solvents. We will continue with the simple finite difference method for the spatial and forward Euler method for temporal dimensions. We will also continue using a one-to-one correspondence between nodal points and pixels when generating the graphics. When we move to 3D, we will adopt a different painting method and use far fewer nodes, which will require some changes to preserve the calculus.

Figure 17. Velocity Field

We will assume a constant porosity of 0.3, though this could also be distributed over the domain in another array. The next thing we will need is a velocity field to represent advection (see figure above). Typical values range from 0.01 to 10 m/day, which roughly correspond to silt and sand, respectively. The longest vector in the figure represents 0.02 m/day. In a real life application, we would use some groundwater flow model to generate this, perhaps FRAC3D[3] or MODFLOW.[4] Here, we will merely fabricate a field, shown as vectors in the preceding figure.

Finite Difference Implementation

We next consider how to implement Equation 2.8 in finite difference form. This expression contains sums $(A+D)$ and products (φC) as well as first and second differentials, making the calculation considerably more complicated. Consider the figure below showing 9 nodes within and surrounding a single differential element or control volume:

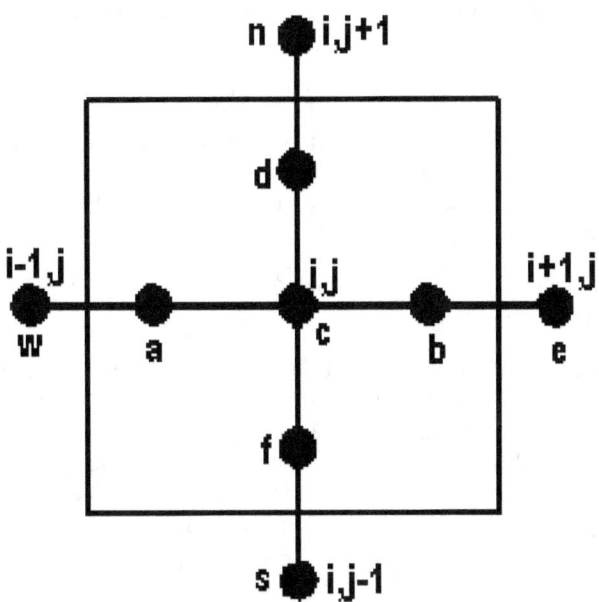

Figure 18. Nodes and Differential Element

The central (i,j), north (i,j+1), south (i,j-1), east (i+1,j), and west (i-1,j) elements are the usual ones for calculating the second derivative and were employed in Chapter 1 (tce2d.c). We have added 4 intermediate nodes labeled a,

---

[3] Therrien, R., and E.A. Sudicky, "Three Dimensional Analysis of Variably–Saturated Fow and Solute Transport in Discretely-Fractured Porous Media," Journal of Contaminant Hydrology, Vol. 23, No. 2, pp. 1-44, 1996.

[4] McDonald, M. G. and Harbaugh, A. W., "*A Modular Three-Dimensional Finite-Difference Ground-Water Flow Model*," USGS Report 83-875, 1984.

b, d, and f. The relationships are simple: a=(w+c)/2, b=(e+c)/2, d=(n+c)/2, and f=(s+c)/2. The first term on the right side of Equation 2.8 becomes:

$$\frac{\partial}{\partial x}\left[(A+D)\frac{\partial(\varphi C)}{\partial x}\right] = \frac{(A+D)_b \frac{\partial(\varphi C)_b}{\partial x} - (A+D)_a \frac{\partial(\varphi C)_a}{\partial x}}{\Delta x} \quad (2.9)$$

The intermediate property values (A_a, A_b, D_a, and D_b) are merely averages:

$$A_a = \frac{A_w + A_c}{2} \quad (2.10)$$

The intermediate partial derivatives are evaluated between two nodes:

$$\frac{\partial(\varphi C)_a}{\partial x} = \frac{\varphi_c C_c - \varphi_w C_w}{\Delta x} \quad (2.11)$$

The Y contributions in the second term on the right side of Equation 2.8 are calculated as for the X contributions with nodes c, d, f, n, and s. Contributions in the third term on the right side of Equation 2.8 are calculated using nodes a, b and d, f, respectively, as in:

$$u\frac{\partial}{\partial x}(\varphi C) = u_c \frac{\varphi_c C_c - \varphi_w C_w}{\Delta x} \quad (2.12)$$

We calculate each of these quantities and then combine them to arrive at the partial derivative with respect to time of the product of porosity and concentration, $\partial\varphi C/\partial t$. Because we kept the porosity, φ, within the parentheses and differentials, this calculation can accommodate spatially varying porosity, which we will implement in the upcoming examples. While it is plausible that porosity might vary over time, that is rarely considered.

$$\frac{\partial(\varphi C)}{\partial t} = \varphi\frac{\partial C}{\partial t} + C\frac{\partial \varphi}{\partial t} \quad (2.13)$$

Upwind Differences

If we were to simply implement the finite difference expressions for advection as for diffusion and dispersion, we would immediately find that the results are unstable. We might decrease the time step and find that the results are still unstable. In fact, no matter how small we make the time step, the results will be unstable. This observation might cause us to revisit the forward Euler method; perhaps replacing it with an implicit or hybrid method, but the results would still be unstable.

If you peruse the literature on advection, you will find many papers devoted to this issue of instability, which comes down to information. If we use a central difference, we are including information (in this case concentration) from where the flow is coming from and also where it is going to but has not yet gotten there; that is, upwind and downwind, respectively. Information downwind is irrelevant or at best noise. In our code implementation, it is truncation errors and round off, which we don't want to include in our calculation; therefore, we take a spatial finite difference on the side upwind of the differential element (e.g., from the west for positive U and from the south for positive V). This scheme was first suggested by Godunov.[5] The entire code can be found in the online archive in folder examples\pce2d in file pce2d.c an excerpt of which is listed below:

```
for(y=1;y<Ny-1;y++) /* interior points */
 {
 for(x=1;x<Nx-1;x++)
   {
   ic=Nx*y+x;
   ie=ic+1;
   iw=ic-1;
   in=ic+Nx;
   is=ic-Nx;
   Cc=C[ic];
   Ce=C[ie];
   Cn=C[in];
   Cs=C[is];
   Cw=C[iw];
   Fc=F[ic];
   Fe=F[ie];
   Fn=F[in];
   Fs=F[is];
   Fw=F[iw];
   Uc=U[ic];
   Vc=V[ic];
   Aa=(Ah[iw]+Ah[ic])/2.;
   Ab=(Ah[ie]+Ah[ic])/2.;
   Ad=(Av[in]+Av[ic])/2.;
   Af=(Av[is]+Av[ic])/2.;
   Ca=(C[iw]+C[ic])/2.;
   Cb=(C[ie]+C[ic])/2.;
   Cd=(C[in]+C[ic])/2.;
   Cf=(C[is]+C[ic])/2.;
   Da=(D[iw]+D[ic])/2.;
   Db=(D[ie]+D[ic])/2.;
   Dd=(D[in]+D[ic])/2.;
   Df=(D[is]+D[ic])/2.;
   dFCdXa=(Fc*Cc-Fw*Cw)/dX;
   dFCdXb=(Fe*Ce-Fc*Cc)/dX;
```

---

[5] Godunov, S. K., "A Difference Scheme for Numerical Solution of Discontinuous Solution of Hydrodynamic Equations",Matematicheskii Sbornik (Transactions of the Moscow Mathematical Society), Vol. 47, pp. 271-306, 1959.

```
dFCdYd=(Fn*Cn-Fc*Cc)/dY;
dFCdYf=(Fc*Cc-Fs*Cs)/dY;
d2ADFCdX2=((Ab+Db)*dFCdXb-(Aa+Da)*dFCdXa)/dX;
d2ADFCdY2=((Ad+Dd)*dFCdYd-(Af+Df)*dFCdYf)/dY;
if(Uc>0.)
   UdFCdX=Uc*(Fc*Cc-Fw*Cw)/dX;
else
   UdFCdX=Uc*(Fe*Ce-Fc*Cc)/dX;
if(Vc>0.)
   VdFCdY=Vc*(Fc*Cc-Fs*Cs)/dY;
else
   VdFCdY=Vc*(Fn*Cn-Fc*Cc)/dY;
dCdt[ic]=(d2ADFCdX2+d2ADFCdY2-UdFCdX-VdFCdY)/Fc;
```

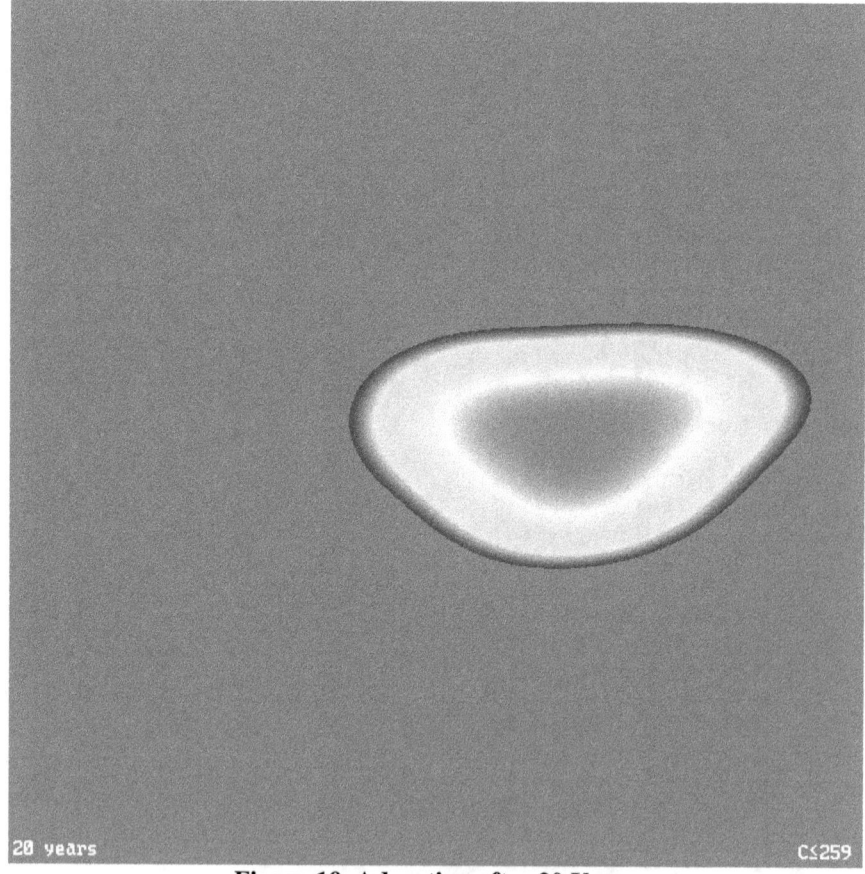

Figure 19. Advection after 20 Years

The solution advances through time as before using the Euler method, only we are now accounting for the spatial variation in properties (diffusion coefficient, dispersion coefficient, and porosity) correctly plus we have added porosity and advection. In just 20 years the center of concentration has moved noticeably to the east as shown in the preceding figure. In 75 years the center of concentration has left the domain:

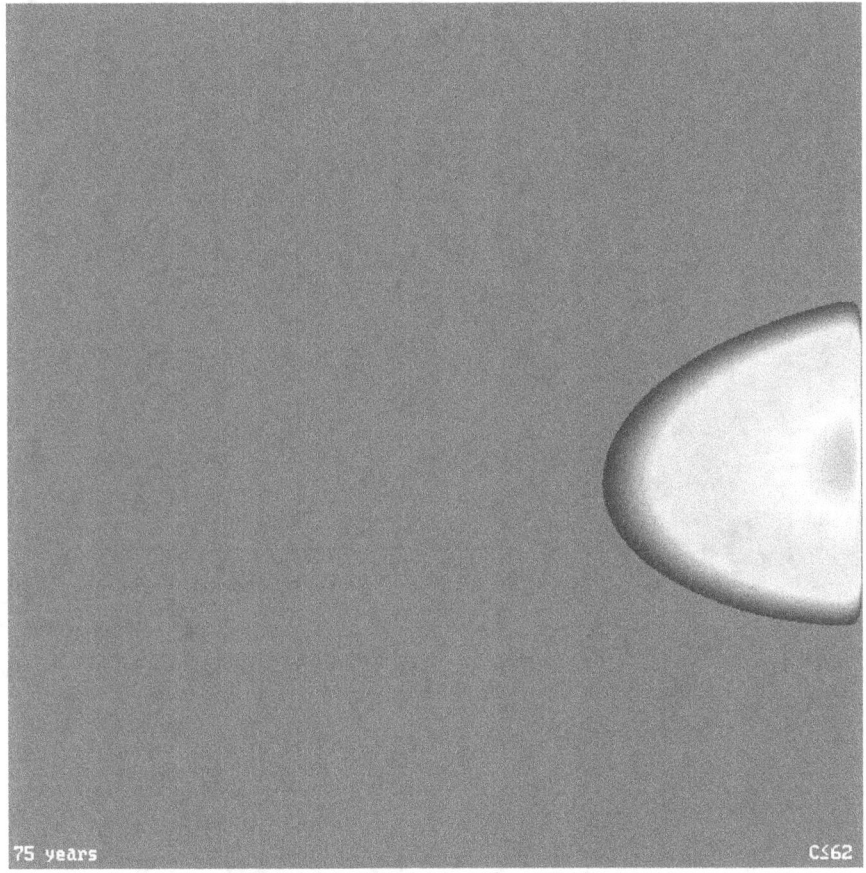

Figure 20. Advection after 75 Years

This domain is only 640m by 640 m, which is rather small compared to the cleanup projects I have worked on, especially the ones created by the DoD. Still, this example does illustrate that, even with a velocity of 0.02 m/day (7.3 m/yr), a contaminant can migrate as well as spread, complicating containment and capture and making remediation difficult.

Some of the largest contaminant plumes I have worked with include those adjacent to Otis Air Force Base and extend over Western Cape Cod, as illustrated in the following figure. These plumes have been remediated. The containment fences of extraction and sampling wells are shown as black circles and plusses, respectively.

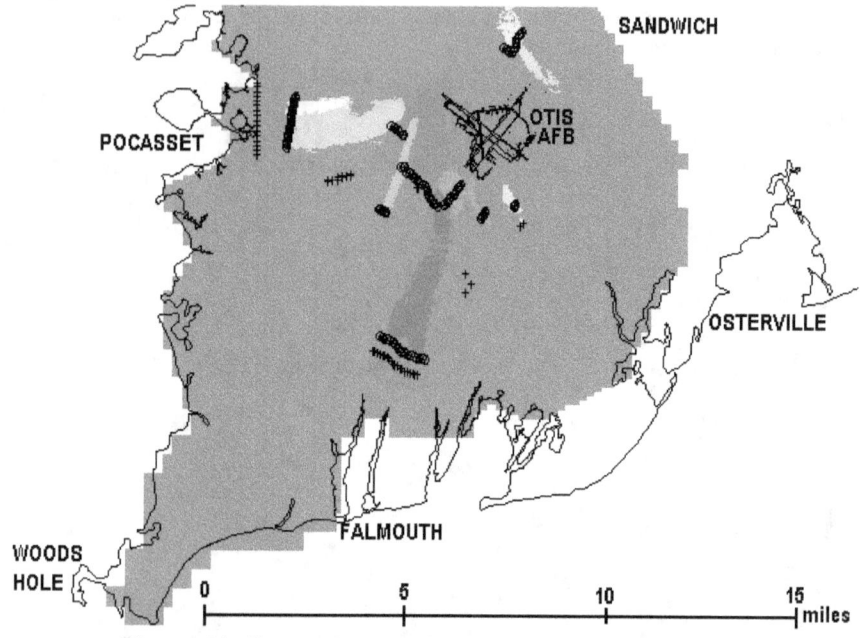

Figure 21. Contaminant Plumes on Western Cape Cod

Spreading and migration of the contaminant can span orders of magnitude, which is why we color the bitmap images in these two example codes (tce2d.c and pce2d.c) based on the log of the concentration.

We can also plot the maximum concentration over time for these various examples, which does somewhat illustrate relative impacts of the available transport mechanisms.

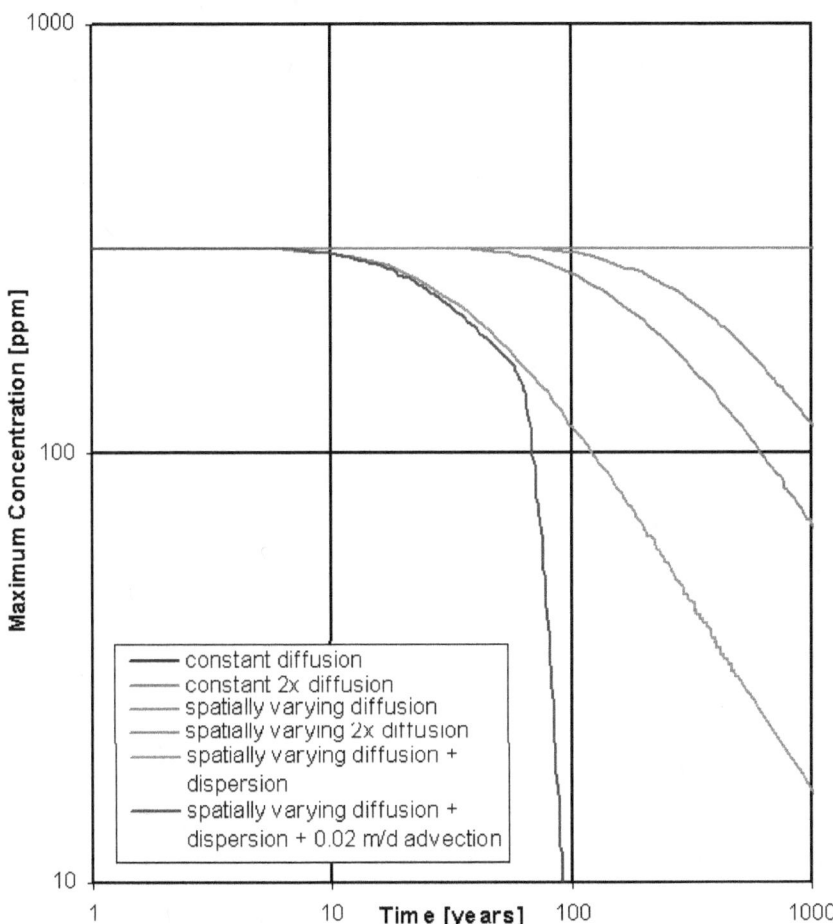

Figure 22. Comparison of 2D Examples

Chapter 4. Contaminant Transport in 3D

The governing partial differential equation that represents the conservation of mass for the contaminant is Equation 2.8 in three dimensions, which is a simple extension from two. While there are many papers and discussions as to which method (e.g., finite difference, finite volume, or finite element) is preferable for modeling fluid flow, those are beyond the scope of this text. The time scale for contaminant transport, whether this be in groundwater, surface water, or atmospheric, is most often much longer than for fluid flow (i.e., water or air). The spatial scale is also much larger. For example, flow through an estuary bordering on the sea will fluctuate somewhat repetitively twice per day because of the tides. There are countless small eddies and separation zones as air flows over a landscape, even a relatively flat one. A contaminant, especially one that will take months—even years—to contain or capture, persists far longer than the finer details of the flowing media in which it resides and with which it migrates. This difference in time scales is why we most often solve such problems in two steps: 1) flow modeling and 2) contaminant transport. We are only concentrating on the second step in this text, as we have covered the first step elsewhere and don't want to detract from the current focus.

Solving the Governing Partial Differential Equation

While the finite volume and finite element methods are indeed useful and may be the most efficient for a particular flow field, these are best suited to obtaining steady state solutions and often involve matrix operations. Peruse the literature on CFD modeling and you will often see references to SOR (successive over-relaxation) and other similar techniques for solving simultaneous linear equations. The version of FRAC3D that I use employs the conjugate gradient method to solve the simultaneous linear equations representing the partial differential equation representing those conservation relationships. In contrast, the finite difference method when applied to contaminant transport begins with an initial condition and steps forward in time. If we had to solve a system of simultaneous linearized equations for each node or integrate over each computational cell applying a Galerkin method at each time step, this would be completely impractical. This is why we use the finite difference method (or particle tracking, as we have discussed elsewhere) to model contaminant transport.

We begin with simple diffusion in 3D, noting that in soil layers, this is often more rapid horizontally than vertically, so we set up our model (tce3d.c in folder examples\tce3d) to handle this. We also consider a different spatial scale (5 km horizontal span 50m vertical). We won't be using a 1m element size, as this would require over two billion nodes. The ratio of discretization in the three directions should be roughly proportional to the rate of change in that direction. If the lateral and horizontal spreading is approximately the same, then Δx and Δy can be the same. If the vertical spreading is approximately 1/10th that in the XY plane, then ∆z should be inversely proportional to that. Note that we must also now consider at least XY and Z separately when selecting a time step. In order to get under one million nodes, we would have to have a grid of ∆x=∆y=25m and ∆z=2.5m or 201x201x21=848,412 nodes. This will take at least twice as long as the 2D models with only 640x640=409,600 nodes.

Estimating Properties

The analytical solution for transient diffusion in one dimension with constant properties was first introduced by Fourier in 1822:

$$C(x,t) = \frac{C_0 \delta}{\sqrt{4\pi Dt}} e^{\left(\frac{-x^2}{4Dt}\right)} \quad (4.1)$$

Holding all other parameters constant and differentiating with respect to t yields the time at which (at a given distance, x) the concentration will be a maximum.

$$t_{max} = \frac{x^2}{2D} \quad (4.2)$$

Plugging this expression back into the preceding equation yields the maximum concentration at this time:

$$C_{max} = \frac{C_0 \delta}{x\sqrt{2\pi}} e^{-0.5} \quad (4.3)$$

These simple analytical expressions can provide an estimate of properties and time scales useful in real world applications. For our first 3D example we will combine diffusion and dispersion, keeping the directions separate. We will arbitrarily select a combined diffusion/dispersion coefficient to spread 1 km in the X direction reaching a maximum in 1800 years and 1/100th this rate in the Y direction. The combined coefficient in the Z direction corresponds to 10 m in this same time frame. While this might seem like an unreasonably long time for a remediation project, it would be reasonable for containment in place, confined by a nearly impermeable cap. In this time we would hope toxic chemicals would break down and radioactive ones decay.

The initial concentration is set to 300 ppm in a "wafer" in the center of the domain having a diameter of 800m and a thickness of 12m. All of the code related to bitmap images has been removed, as we are not painting the results, rather writing them out to sequential files. While we could put multiple "snapshots" (i.e., representations of the concentration field at a particular time) in the same file, making it 4D, these would be so large as to make loading them all into memory at the same time problematic and likely beyond the available physical memory. Tecplot™ can accept multiple "zones" which can be turned on and off to produce time series animations.

We have also simplified the allocation of memory within the program plus added functions to find the minimum or maximum value in an array. These facilitate calculation of the time step and also reporting of the maximum concentration. The finite difference equation has also been expanded to incorporate the third (vertical) dimension and the natural boundary conditions have been added for the bottom and top. The log of the initial concentration is shown in the following figure:

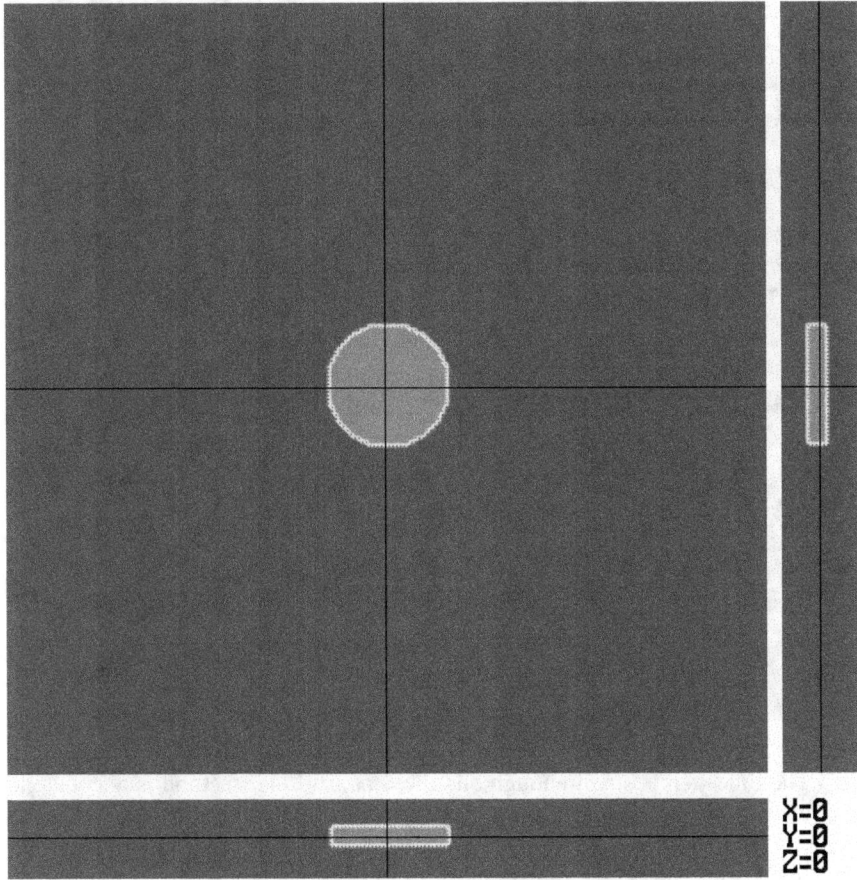

Figure 23. Initial log Concentrations in 3D

We must next consider the time step required to obtain a stable solution. To calculate this, we find the largest value of the combined diffusion/dispersion coefficient in each of the three directions, divide this into the square of the grid spacing in that dimension and take half (e.g.. $\Delta t \leq \Delta x^2/2D_{max}$). This yields 1.1, 113, and 113 years for the X, Y, and Z dimensions, respectively. As we don't want to step forward more than one year for generating output, we round this down to ∆t=1 yr. The results after 100 years are shown in this next figure:

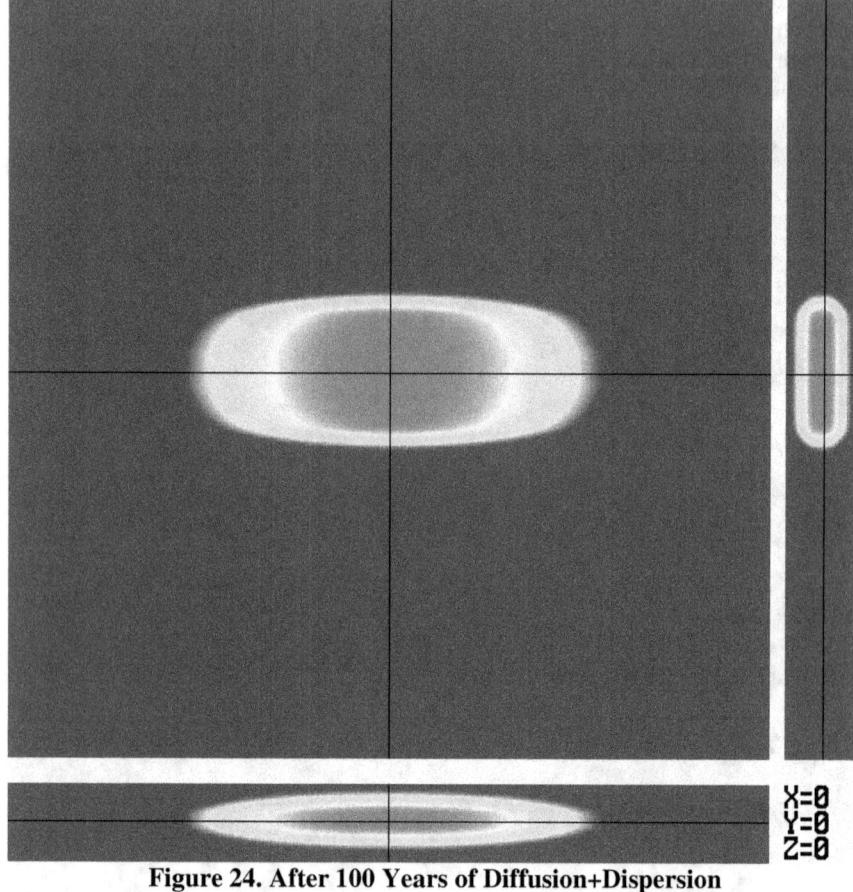

Figure 24. After 100 Years of Diffusion+Dispersion

The time step is small enough and the forward Euler method is adequate to provide a slow but stable solution. We march through time and see that the numerical solution in 3D roughly corresponds to the magnitude of spreading we presumed when selecting the dimensions and properties. This observation is important in that there are many steps in a successful remediation project. Rough estimates are quite useful when evaluating alternatives, such as containment vs. removal (e.g., pump-and-treat). The cost of such options may vary greatly. The time required to implement different options may also vary greatly. These considerations and more should guide the remediation design process if a successful outcome is to be achieved. The concentrations after 1800 years are shown in the following figure:

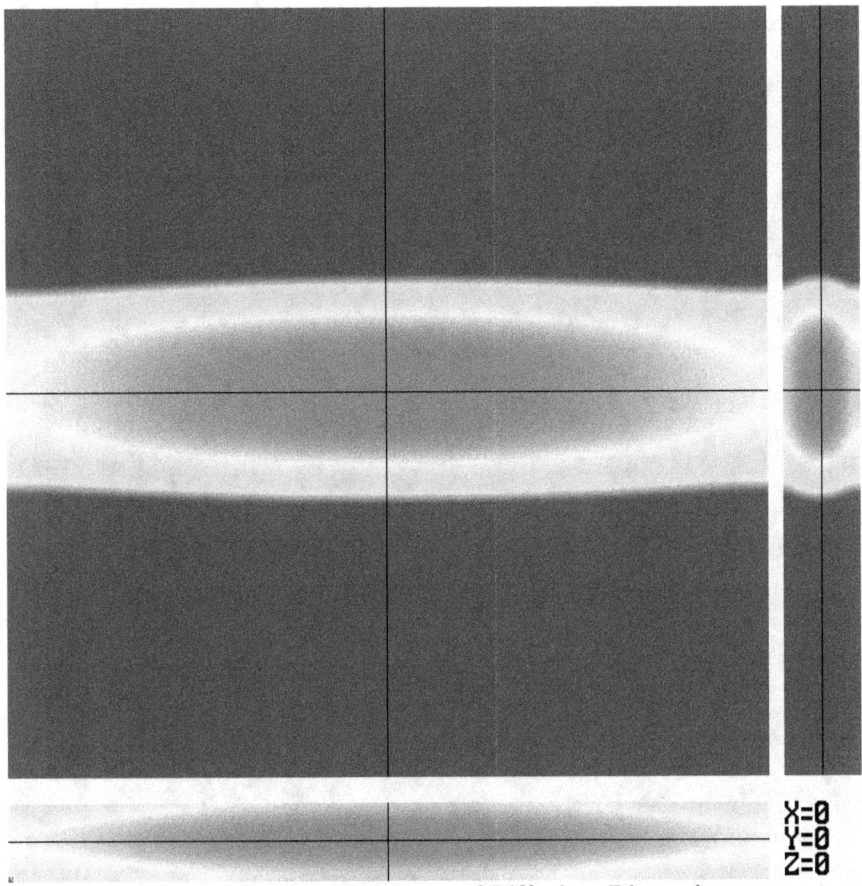

Figure 25. After 1800 Years of Diffusion+Dispersion

Advection

Advection will be present in most sites unless these have been specially prepared to contain the contaminant. We add the same field as before (Figure 17) with a much smaller vertical component and again switch folders and file names to pce3d so as to not overly complicate the original example. As for the 2D example, at this step we also add porosity.

As with 2D advection, we use upwind differences. With the third dimension, we must also add nodes to the arrangement in Figure 18, including an upper and lower plus two more vertical intermediate points, designated g and h. The time step must now consider all transport mechanisms (e.g., $\triangle x^2/2D$ and also $\triangle x/6U$), taking the most restrictive (i.e., smallest) one to avoid instability. After 100 years of combined diffusion/dispersion and advection (approximately

0.02 m/day) the contaminant has spread, migrated to the right and distorted in shape, as shown in this next figure:

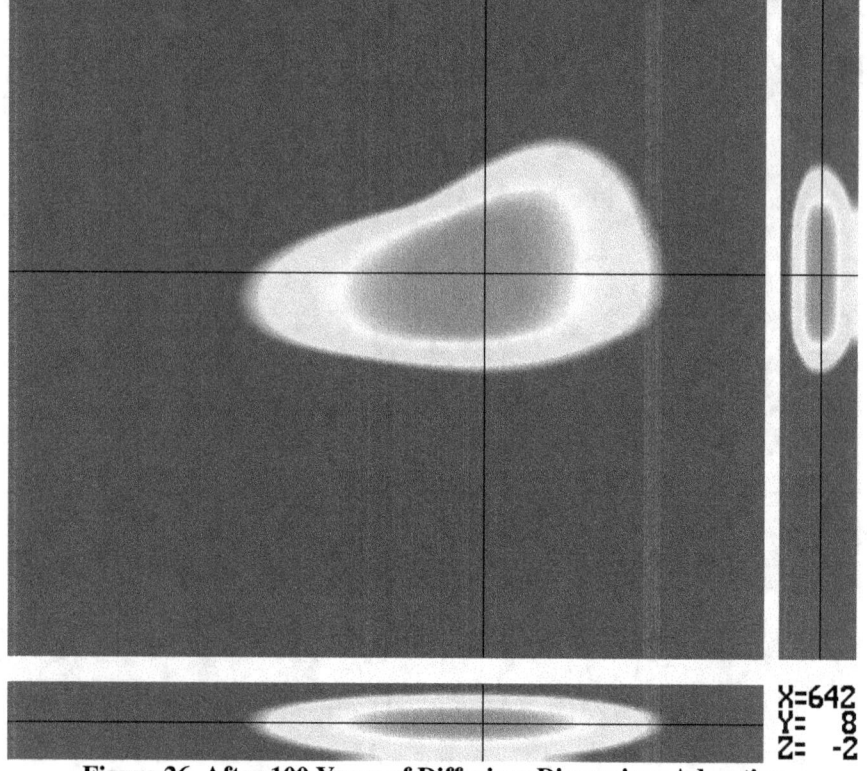

Figure 26. After 100 Years of Diffusion+Dispersion+Advection

The contaminant plume is responding as we would expect, given the properties and flow we have applied to it. It is spreading more laterally in the X direction than horizontally in the Y direction. It is also spreading more downward than upward vertically. We are also starting to see some rotational distortion due to the curling velocity field. After 250 years, the plume has taken on a droplet shape and the center of mass has move to approximately X=1200m, Y=-25m, Z=-5m. We also see more clockwise rotational distortion. The colors (representing the log of the concentration) smoothly change from blue to cyan to green to yellow to orange and finally to red as we approach the center of mass and backward as we move away from the center of mass in any of the three directions. This is evidence that the time stepping (forward Euler method) is still working adequately.

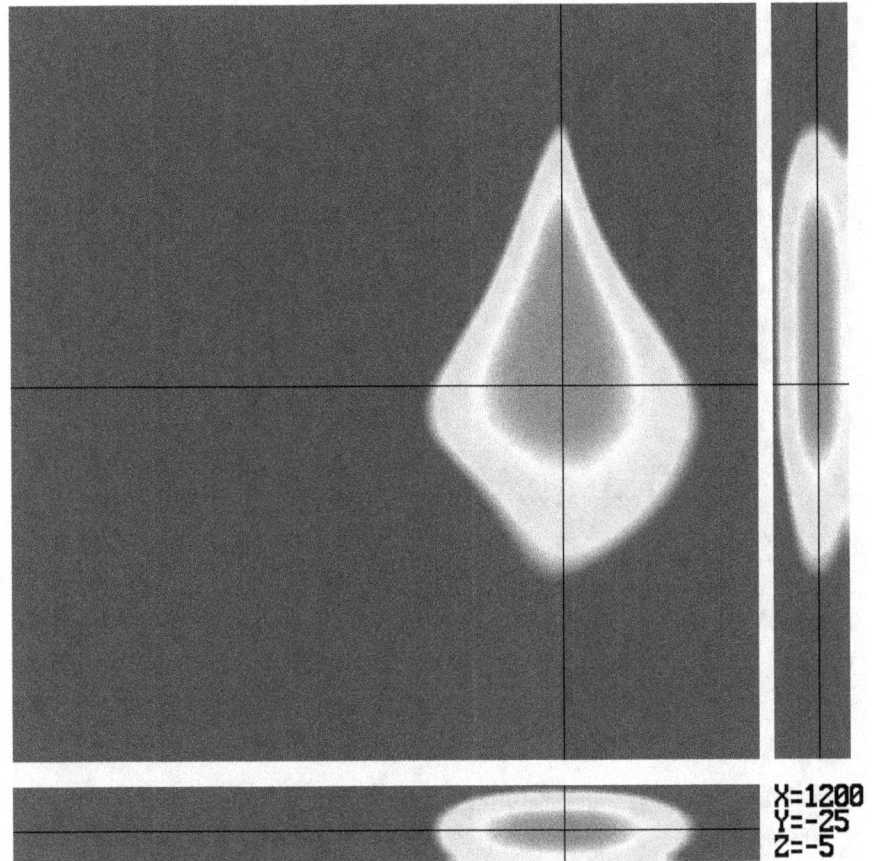

Figure 27. After 250 Years of Diffusion+Dispersion+Advection

This plume is still moving considerably slower than some I have modeled, including those on Western Cape Cod shown in Figure 21, which moved a distance on the order of a kilometer in approximately 25 years. The soil there has a higher hydraulic conductivity and porosity than some other locations within the continental US. In the next example we will adjust the magnitude and also pattern of the flow to see how this impacts the plume.

After 500 years the plume has distorted even more, as seen in the following figure and is even starting to leave the boundary on the right side and also the bottom. We can also see the natural boundary conditions in this figure. The gradients (spatial slope vectors) or directional rate of change of the concentrations, as indicated by the color variation, are smooth at the edges, rather than stacking up colors at the edge.

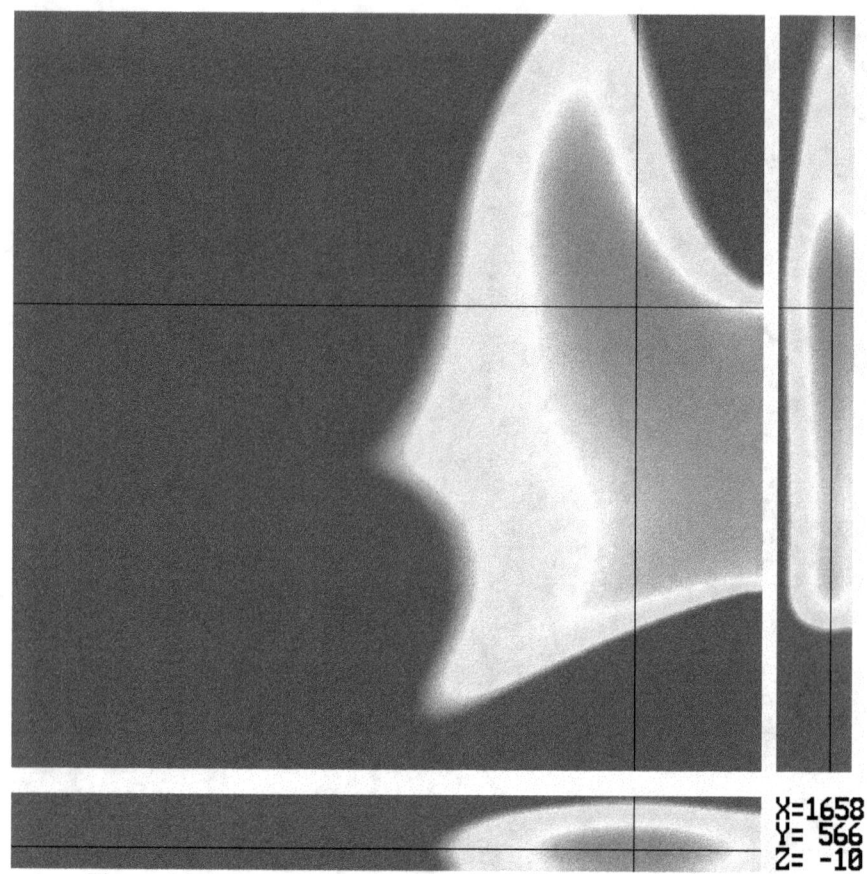

Figure 28. After 500 Years of Diffusion+Dispersion+Advection

Chapter 5. MODFLOW Based Models

The preceding examples have served to illustrate the principles of modeling contaminant transport but are not well suited to actual problems. We need code that will handle a variety of problems conveniently and not require editing and recompilation for each one. That means reading input files, particularly those that have been generated by some other software, such as Build3D (see Appendix D). We will consider two different flow models: MODFLOW in this chapter and FRAC3D in the next. We will read the respective input files (nodes, elements, and properties) and output files (flows) plus a file containing initial concentrations and then march through time calculating the contaminant transport. We will not dwell on the details of MODFLOW in this text. The reader is directed to the many resources available for more specific information.

To accomplish this, we will use an example (ctmod.c), which can be found in the online archive in the examples\modflow folder along with several sets of input files. The output files generated will be compatible with TP2. Should you prefer output compatible with Tecplot™, swap out the function with the code (ctfra.c) discussed in the next chapter.

The prefix for the first example we will consider in this chapter is WINE. Launch the program (ctmod.exe) passing it the parameter WINE. Several files must exist and be in the same folder. The examples are in subfolders. You can either copy the files into the folder with the executable or copy the executable into each of the example folders or go to one of the example folders and type something like the following to launch the executable from the parent folder:

```
..\CTMOD WINE
```

The first four files read for this example will be WINE.BAS (the MODFLOW basic input file), WINE.BCF (the block-centered file), WINE.CBB (the binary flow flow), and WINE.ELP (the property file). This is what you should see:

```
3D Transient Diffusion + Dispersion + Advection
reading MODFLOW input and output files
basic input file: WINE.BAS
  title: PREFIX: WINE
  subtitle: CREATED BY BUILD3D/V1.71
  grid: 45x37x5
  total cells: 8325
  active cells: 5860
  element type: hexahedra (bricks)
  augmented grid: 46x38x6
  resulting nodes: 10488
  active nodes: 7536
block-centered file: WINE.BCF
  0≤X≤7889.85
  0≤Y≤6500.16
  0≤Z≤235.105
```

```
binary flow file: WINE.CBB
    -41.0391≤U≤53.6929
    -27.8412≤V≤40.8043
    -3.3188≤W≤12.3406
element property file: WINE.ELP
    0.3≤F≤0.3
    3≤Dx≤3
    3≤Dy≤3
    0.3≤Dz≤0.3
```

The last file read contains the initial concentrations, WINE.ICC. Note that MODFLOW begins with a full grid (i.e., a block of elements). Some elements may be inactive so as to create an irregular domain. Initial concentrations for this example are shown in the following figure:

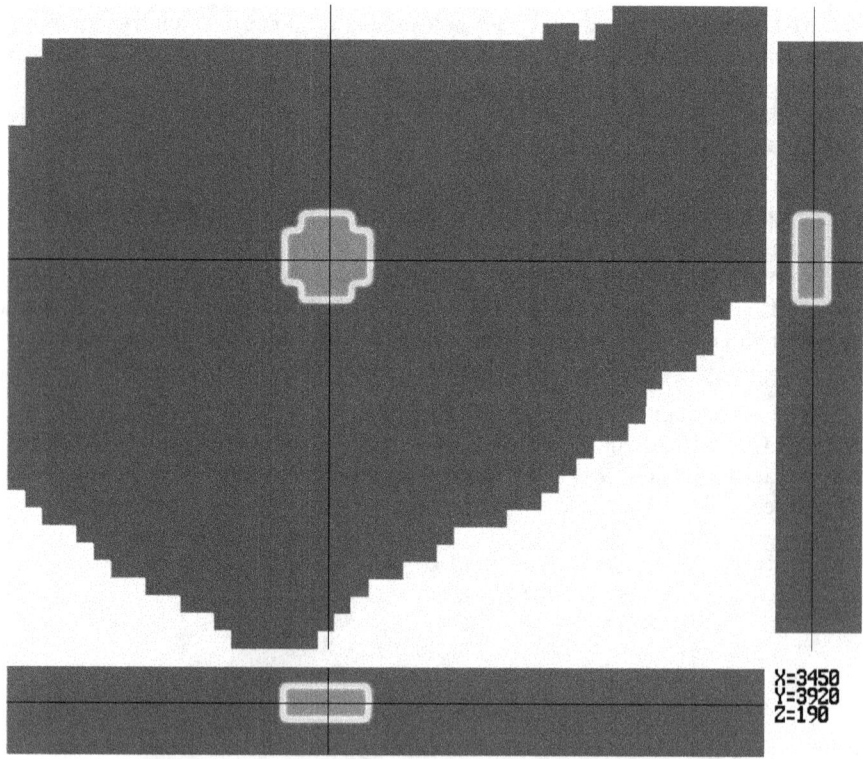

Figure 29. Initial Concentrations for WINE Example

This figure also shows the active and inactive parts of the domain (blue vs. white). As MODFLOW is element-based and ctmod is node-based, the initial concentrations have been relocated to the nearest node (refer to function ReadConcentrations() in source code file ctmod.c).

The vertically-averaged velocity vectors are shown in this next figure:

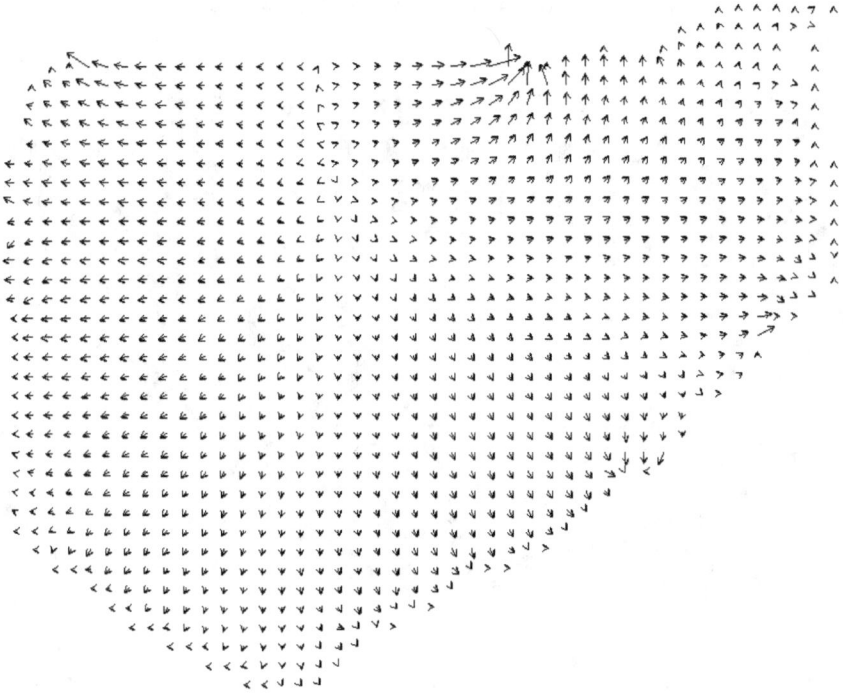

Figure 30. Vertically-Averaged Velocity Vectors for WINE Example

As before with 3D advection (see pce3d.c), we must consider several things when selecting a time step. A summary of these is listed below:

```
determining time step
   ΔX=175.33,  ΔY=175.68,  ΔZ=14.81
   ΔX²/Dx/2=5123.43,  ΔY²/Dy/2=5143.91,  ΔZ²/Dz/2=365.56
   ΔX/U/6=0.544237,  ΔY/V/6=0.717572,  ΔZ/W/6=0.200017
   Δt=0.121452
```

We advance the solution as before, only working from the elements so that we can only consider those that are active. MODFLOW defines properties within the elements. Note that all of the properties must be consistent. It doesn't matter what properties are used (English or metric), as long as they are consistent. The velocities must in length/time and the diffusion/dispersion coefficients must be in length²/time. The time step is dX^2/D, whatever that works out to be. The same is true for the concentrations. The output will have the same units as the input.

After 25 time units (presumably years), the maximum concentration has dropped from 300 to 285 and spread considerably, as shown in the following figure:

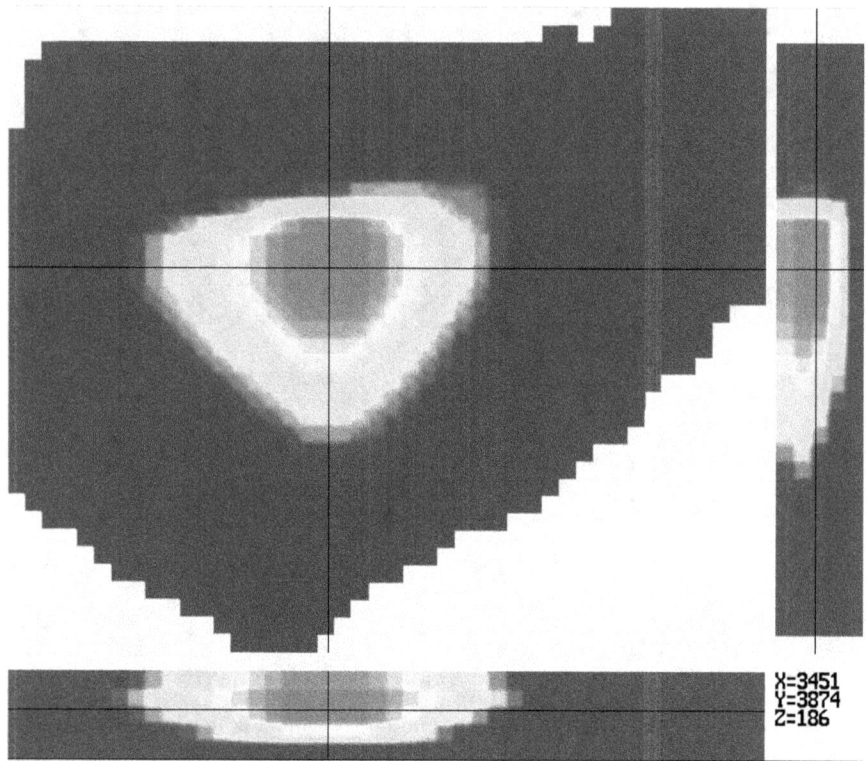

Figure 31. WINE Concentrations (log) after 25 Years

After 100 years the plume has spread even further and noticeably migrated laterally. The impact of the velocity field is also clear after this period.

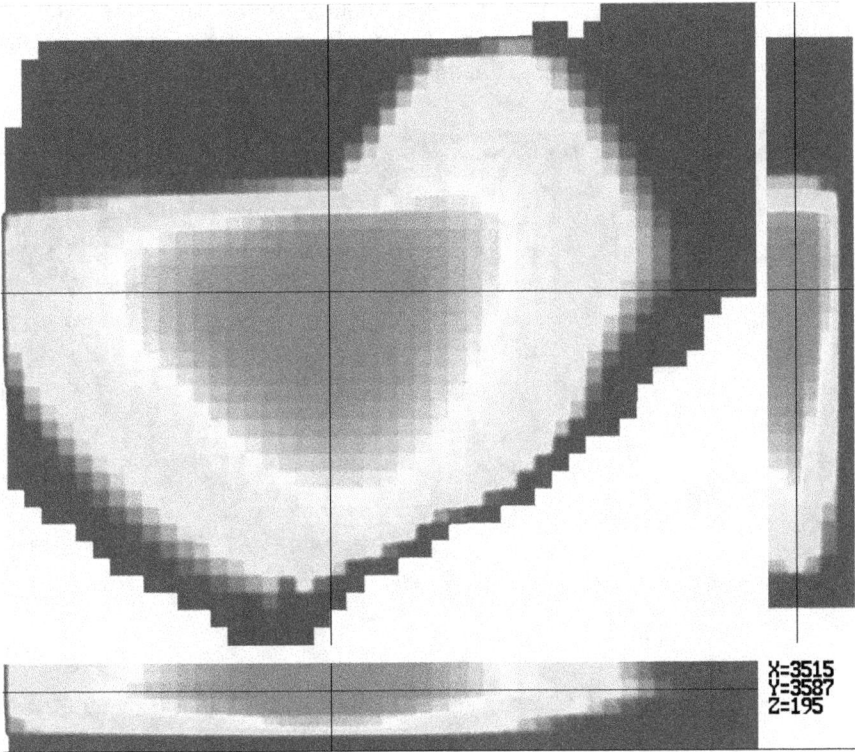

Figure 32. WINE Concentrations (log) after 100 Years

We end the simulation at 650 years, as the maximum concentration has decreased by a factor of 10 (an arbitrary number in this case). This would definitely be consider a complete failure if containment or capture and removal of the contaminant were the objective. This particular scenario is the "no action" alternative, which was not chosen in this case.

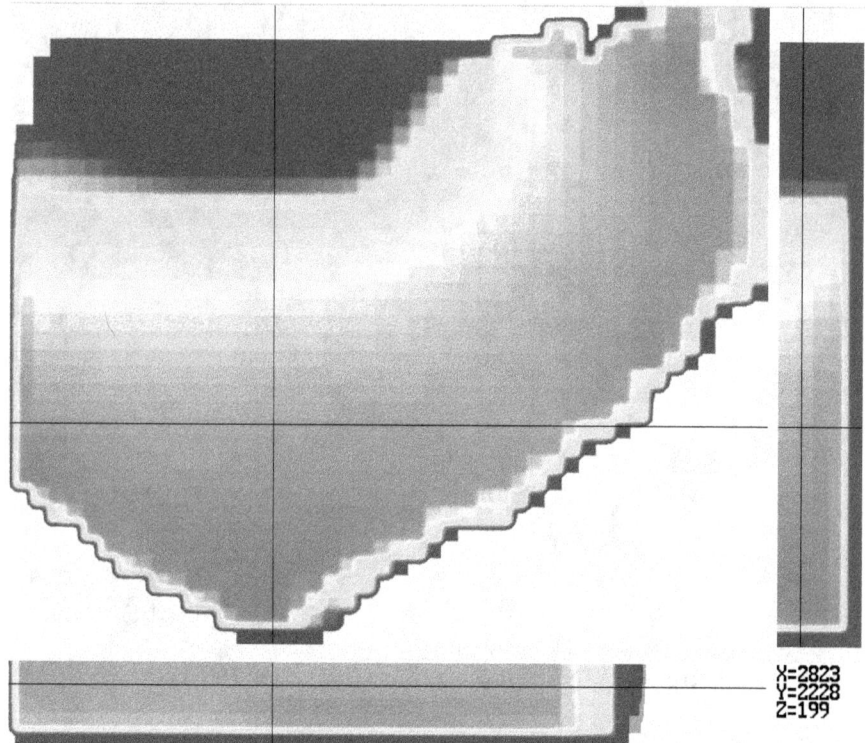

Figure 33. WINE Concentrations (log) after 650 Years

Multiple Plumes and Contaminants

Some remediation projects contain multiple contaminated sites, like where some guy emptied a tanker truck into a vacant lot. [I am not making this up!] Some plumes contain multiple contaminants. This is one reason for using particle tracking instead of the type of transport models we have discussed so far. PTRAX can handle up to 128 different plumes simultaneously with up to 128 different contaminants in each plume. There are 10 different plumes shown in Figure 21. You will find a separate "seed" file for each plume in the folder examples\MODFLOW\OTIS (see files OTIS?.SED). These each contain only a few particle seeds. The full model contained tens of thousands of seeds but those files are quite large and so are not contained in the online archive of examples.

We begin this next MODFLOW-based example (OTIS) with three contaminated spots, as shown in the following figure:

Figure 34. Initial Concentrations OTIS Example OTIS

The vertically-averaged velocity field is shown in this next figure:

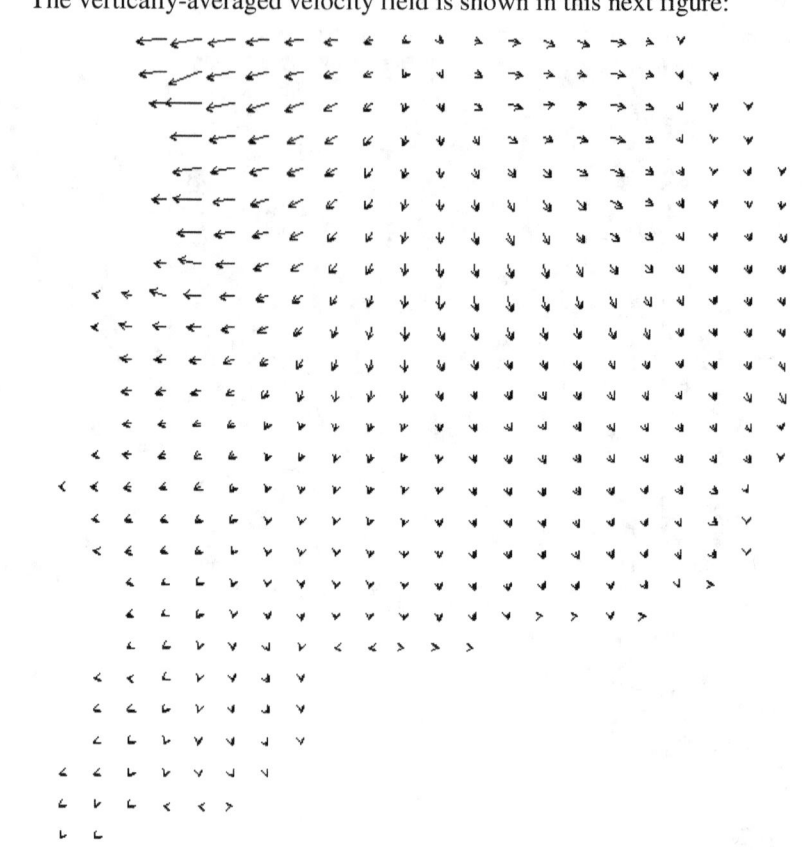

Figure 35. Vertically-Averaged Velocity Vectors for OTIS Example

Launching the program produces the following output:

```
ctmod otis
3D Transient Diffusion + Dispersion + Advection
reading MODFLOW input and output files
prefix: otis
basic input file: OTIS.BAS
  title: PREFIX: OTIS
  subtitle: CREATED BY BUILD3D/V1.73
  Reference Point: X=825900, Y=189450
  grid: 25x29x5
  total cells: 3625
  active cells: 2150
  augmented grid: 26x30x6
  element type: hexahedra (bricks)
  resulting nodes: 4680
  active nodes: 2946
block-centered file: OTIS.BCF
  825900≤X≤887400
  189450≤Y≤255350
  -347.58≤Z≤65.055
element property file: OTIS.ELP
  0.3≤F≤0.3
  3≤Dx≤3
  3≤Dy≤3
  0.3≤Dz≤0.3
binary flow file: OTIS.CBB
  -260.426≤U≤112.026
  -143.778≤V≤31.8607
  -1.65294≤W≤4.03023
vertically-averaged velocities: OTIS.V2D
initial concentrations: OTIS.ICC
determining time step
  ΔX=2460, ΔY=2272.41, ΔZ=21.09
  ΔX²/Dx/2=1.0086E+006, ΔY²/Dy/2=860641,
ΔZ²/Dz/2=741.313
  ΔX/U/6=1.57435, ΔY/V/6=2.63416, ΔZ/W/6=0.872159
  Δt=0.462376
0 years C≤300 OTIS000.TB3
25 years C≤290 OTIS001.TB3
50 years C≤264 OTIS002.TB3
75 years C≤238 OTIS003.TB3
100 years C≤211 OTIS004.TB3
```

We can see some spreading but no discernable migration after 100 years:

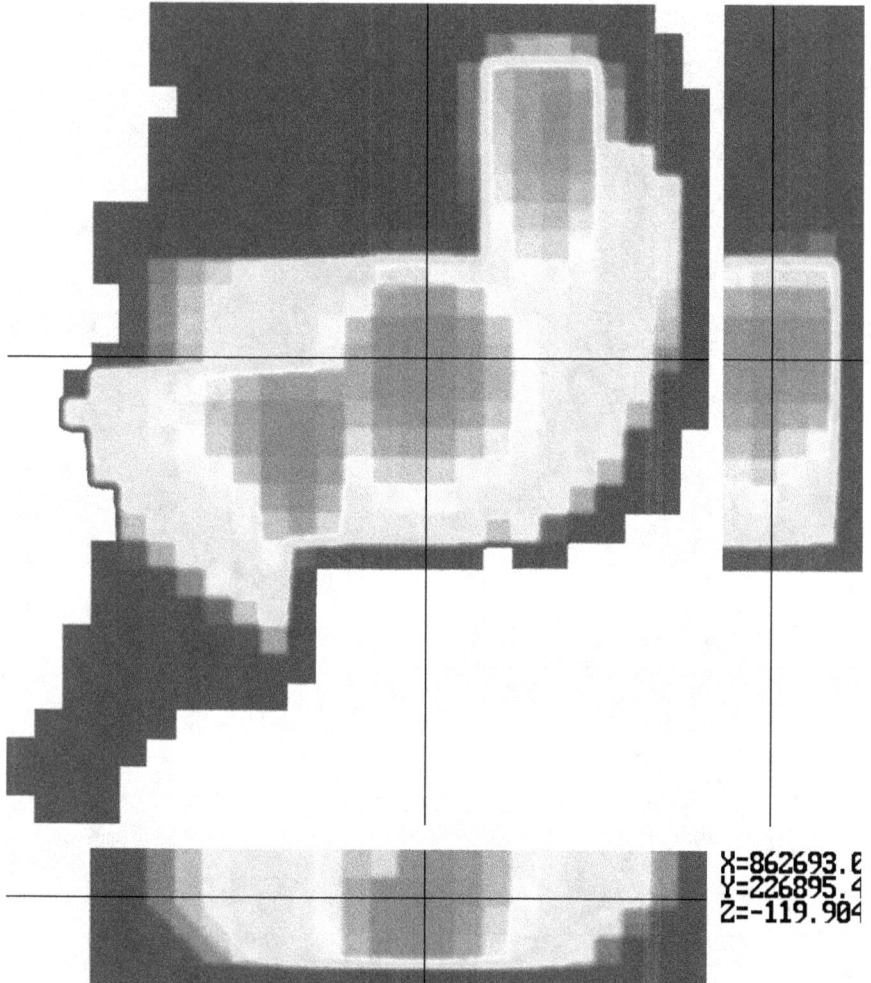

Figure 36. OTIS Concentrations (log) after 100 Years

After 500 years the contaminants have spread considerably and moved appreciably.

Figure 37. OTIS Concentrations (log) after 500 Years

Impact of Surface Slope

This next example is a valley with considerable surface slope illustrated in this 3D perspective view:

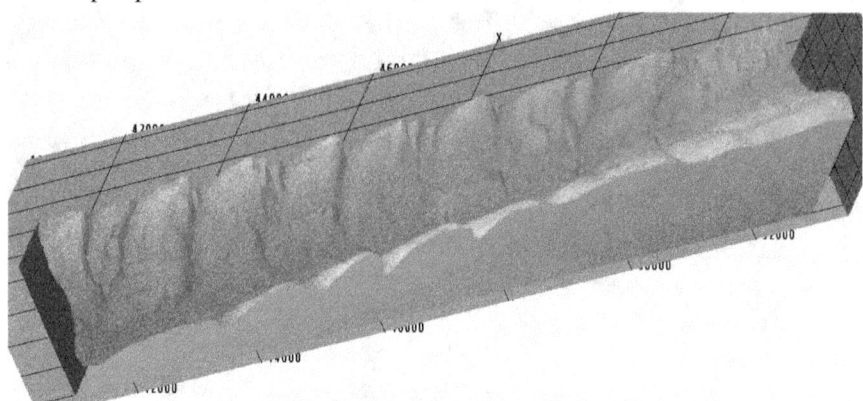

Figure 38. Perspective View of BEAR Example

The surface slope influences the groundwater velocities, as shown in this next figure:

Figure 39. Vertically-Averaged Velocities for BEAR Example

There were numerous contaminants at this remediation site, which was a burial ground for radioactive, toxic, and otherwise hazardous materials, only a few miles from my house. Dozens of plumes were tracked and several different types of capture and removal approaches were taken to clean up this site. For the sake of simplicity, we will consider a single plume emanating from a spot near the top of the valley.

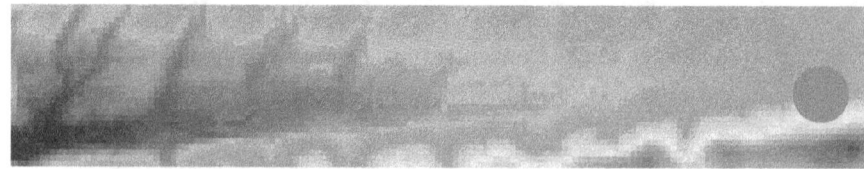

Figure 40. Contaminant Source Location for BEAR Example

Launching the program produces the following output:
```
ctmod bear
3D Transient Diffusion + Dispersion + Advection
reading MODFLOW input and output files
```

```
prefix: bear
basic input file: BEAR.BAS
  title: PREFIX: BEAR
  subtitle: CREATED BY BUILD3D/V1.71
  grid: 60x15x5
  total cells: 4500
  active cells: 4500
  augmented grid: 61x16x6
  element type: hexahedra (bricks)
  resulting nodes: 5856
  active nodes: 5856
block-centered file: BEAR.BCF
  0≤X≤11925
  0≤Y≤2224.95
  0≤Z≤1189.1
element property file: BEAR.ELP
  0.3≤F≤0.3
  3≤Dx≤3
  3≤Dy≤3
  0.3≤Dz≤0.3
binary flow file: BEAR.CBB
  -5975.29≤U≤591.701
  -3032.46≤V≤1861.64
  -1788.07≤W≤1160.5
vertically-averaged velocities: BEAR.V2D
initial concentrations: BEAR.ICC
determining time step
  ΔX=198.75, ΔY=148.33, ΔZ=160.45
  ΔX²/Dx/2=6583.59, ΔY²/Dy/2=3666.96, ΔZ²/Dz/2=42907
  ΔX/U/6=0.00554366, ΔY/V/6=0.00815234, ΔZ/W/6=0.0149556
  Δt=0.00270332
  0 years C≤300 BEAR000.TB3
 25 years C≤300 BEAR001.TB3
 50 years C≤299 BEAR002.TB3
 75 years C≤298 BEAR003.TB3
100 years C≤298 BEAR004.TB3
125 years C≤297 BEAR005.TB3
150 years C≤297 BEAR006.TB3
175 years C≤296 BEAR007.TB3
200 years C≤296 BEAR008.TB3
225 years C≤295 BEAR009.TB3
250 years C≤294 BEAR010.TB3
275 years C≤294 BEAR011.TB3
300 years C≤293 BEAR012.TB3
325 years C≤292 BEAR013.TB3
350 years C≤292 BEAR014.TB3
375 years C≤291 BEAR015.TB3
400 years C≤291 BEAR016.TB3
425 years C≤290 BEAR017.TB3
```

```
450 years  C≤289  BEAR018.TB3
475 years  C≤289  BEAR019.TB3
500 years  C≤288  BEAR020.TB3
```

After 100 years we see that the contaminant has spread down hill for almost the entire length of the valley, yet there is still a significant concentration at the source. This is characteristic of some sites that have significantly different properties and/or velocities in the three different spatial directions (horizontally, laterally, and vertically).

Figure 41. BEAR Concentrations (log) after 100 Years

We see very little change between 100 and 500 years:

Figure 42. BEAR Concentrations (log) after 500 Years

The MADE Site

The last MODFLOW-based example we will consider comes from the MADE site, for which I provided modeling support and software development for the first decade. We learned much from this site, not only about the soil properties and interactions with the contaminants, but also about the software required to manage the data collected and build models. The 3D hydraulic conductivity field was one of the most complex tasks I tackled early on. This next figure shows the original hydraulic conductivity field for the entire site.

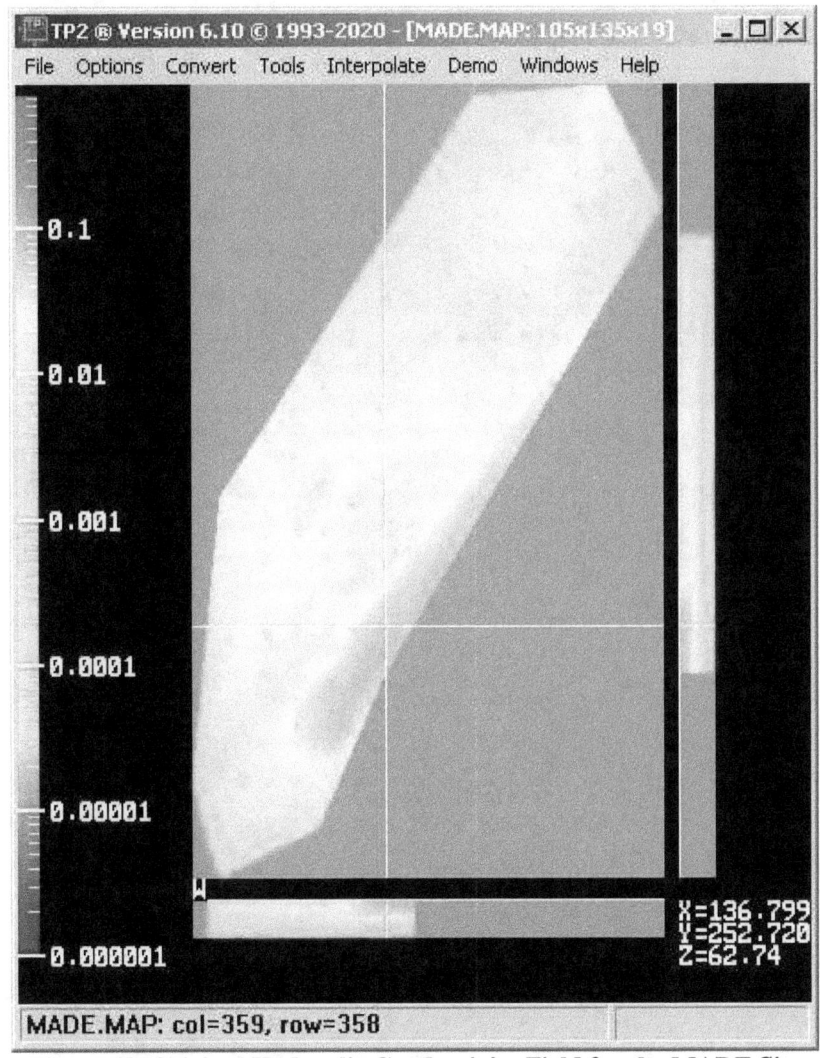

Figure 43. Original Hydraulic Conductivity Field for the MADE Site

I developed a program (Field3D) to handle this data, as nothing was available commercially at that time capable of dealing with it. This functionality was later built into TP2. More recent versions (7.04 and later) of Tecplot™ can also handle this type of data. Many studies were conducted at the MADE site and sections of the entire area were the subject of specific focus, including the cutout region in the MADE example that can be found in the online archive in the examples\MODFLOW\MADE folder. The vertically-averaged velocity vectors for this cutout region are shown in the following figure:

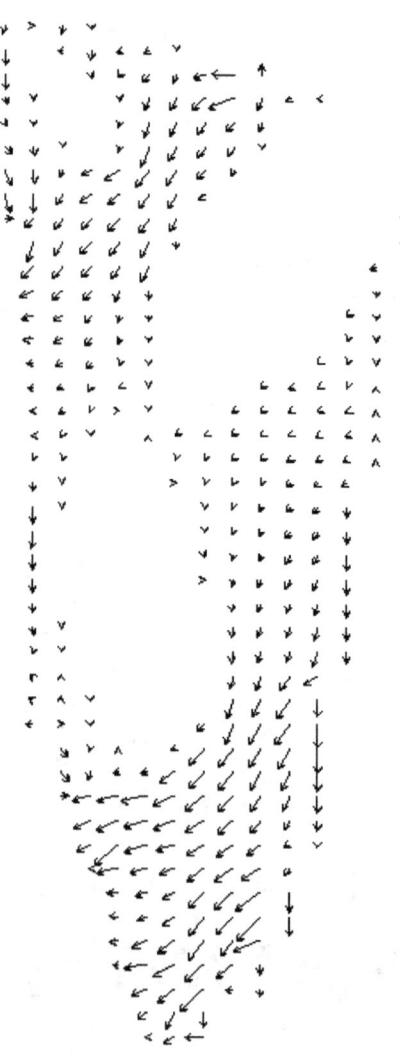

Figure 44. Vertically-Averaged Velocity Vectors for MADE Example

While many plumes were created, we will only consider a few in this example. The initial concentrations are shown in this next figure:

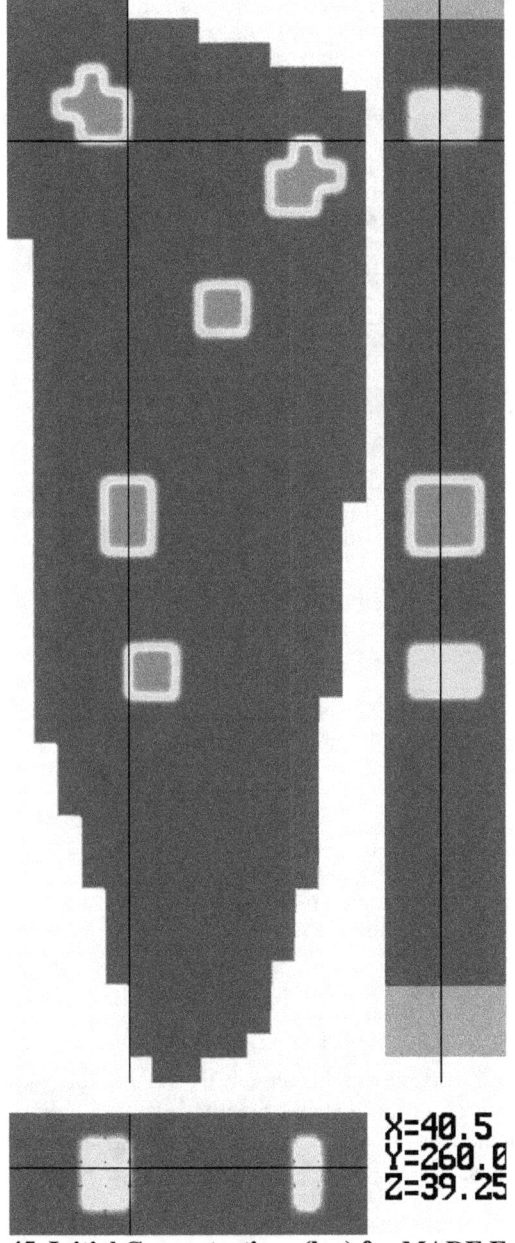

Figure 45. Initial Concentrations (log) for MADE Example

Launching the program produces the following output:

```
ctmod made
3D Transient Diffusion + Dispersion + Advection
reading MODFLOW input and output files
prefix: made
basic input file: MADE.BAS
   title: PREFIX: MADE
   subtitle: CREATED BY BUILD3D/V1.71
   grid: 15x45x5
   total cells: 3375
   active cells: 2350
   augmented grid: 16x46x6
   element type: hexahedra (bricks)
   resulting nodes: 4416
   active nodes: 3180
block-centered file: MADE.BCF
   0≤X≤120
   0≤Y≤300.001
   0≤Z≤118.205
element property file: MADE.ELP
   0.3≤F≤0.3
   3≤Dx≤3
   3≤Dy≤3
   0.3≤Dz≤0.3
binary flow file: MADE.CBB
   -28.558≤U≤13.8079
   -30.3216≤V≤10.8248
   -9.41612≤W≤11.996
vertically-averaged velocities: MADE.V2D
initial concentrations: MADE.ICC
determining time step
   ΔX=8, ΔY=6.6667, ΔZ=14.023
   ΔX²/Dx/2=10.6667, ΔY²/Dy/2=7.40748, ΔZ²/Dz/2=327.741
   ΔX/U/6=0.0466885, ΔY/V/6=0.0366444, ΔZ/W/6=0.194829
   Δt=0.0184937
 0 years C≤300 MADE000.TB3
25 years C≤91 MADE001.TB3
50 years C≤50 MADE002.TB3
75 years C≤36 MADE003.TB3
100 years C≤29 MADE004.TB3
```

The MADE site was quite small compared to the other cleanup projects the Team worked on. It was chosen because it was inside the guarded perimeter of the Air Force Base and wouldn't be disturbed. It also provided a much faster reaction time, which was more conducive to study than the much longer remediation projects.

The concentrations after 25 years are shown in the figure below. This was a small site compared to

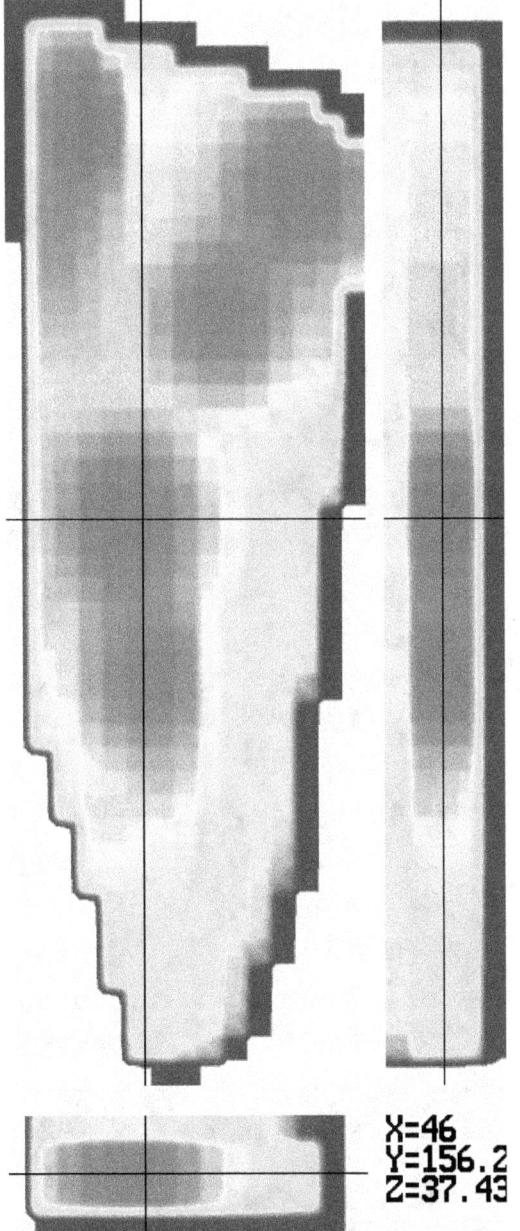

Figure 46. MADE Concentrations (log) after 25 Years

The concentrations after 100 years are shown in the following figure:

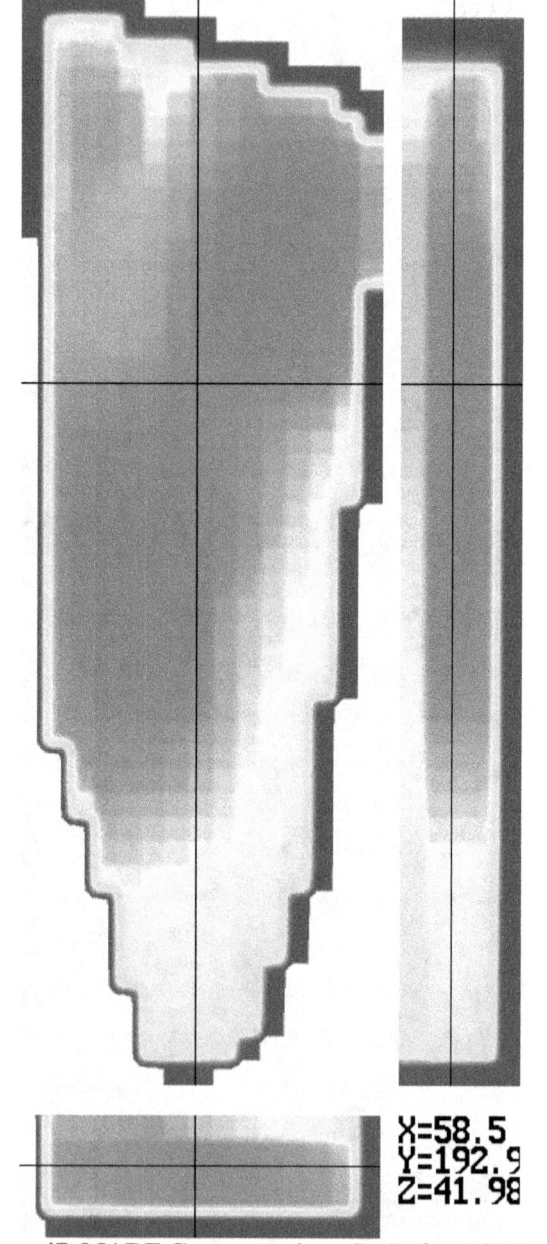

Figure 47. MADE Concentrations (log) after 100 Years

Chapter 6. FRAC3D Based Models

We will now modify the code from the previous chapter to read FRAC3D input and output files, which are very different from MODFLOW files. The code (ctfrac.c), along with several examples in this second format, can be found in the online archive in the examples\frac3d folder. The output files generated will be compatible with Tecplot™. Should you prefer output compatible with TP2, swap out the function with the code (ctmod.c) discussed in the previous chapter or convert the files, as described in the corresponding appendices.

The MODFLOW-based examples in the previous chapter began with a full grid with some cells deactivated in order to create irregular domains. FRAC3D works differently by reading nodes and elements, which are connected by virtue of common nodes. While the MODFLOW models all had regularly-spaced grids, this is not necessarily the case with FRAC3D models. Lacking this regular pattern of nodes and elements, we must figure out which nodes are connected and how. As shown in the following figure, with stacked hexahedral elements, a node may have up to 26 neighbors.

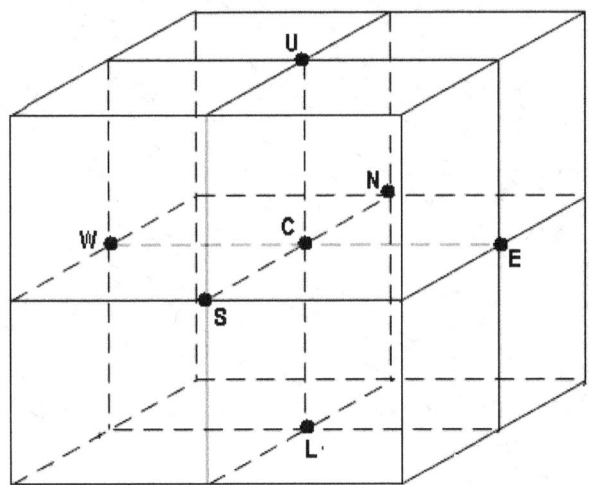

Figure 48. Stacked Hexahedra (Bricks) Showing Adjacent Nodes

We first build a list of adjacent nodes. [It is assumed here that the reader understands how to loop through elements associating each pair of nodes. If not, refer to ctfra.c function NodeNeighbors().] If a node doesn't have 26 neighbors, then it is on a boundary. The natural boundary conditions we applied previously were the partial derivatives of the concentration along a direction perpendicular to any boundary was assumed to be zero. This same natural boundary condition will result with the current implementation, as the finite difference contribution in that direction (toward the boundary) will be zero. We must also move the properties from the elements to the nodes, which is a simple averaging process.

The properties (Dx, Dy, Dz, and φ) for each node will be the average of all the elements containing this node.

The elements for the first FRAC3D-based example (LAFB) are shown in the following figure:

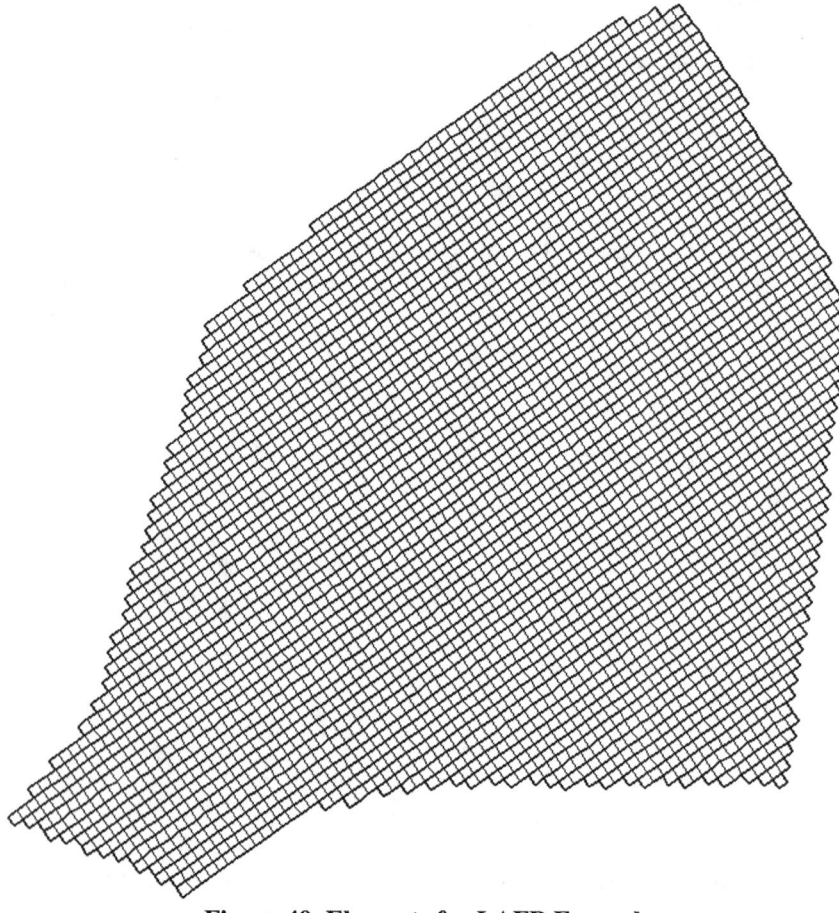

Figure 49. Elements for LAFB Example

The element outlines can be found in file LAFB.P2D and can be drawn with TP2. The nodes are in LAFB.NDE, the elements are in LAFB.ELM, the properties are in LAFB.PRP, the (porous media) velocities are in LAFB.VEP, and the initial concentrations in LAFB.ICN.

The velocity vectors are written to LAFB.VEC formatted for Tecplot™ and are shown in this next figure:

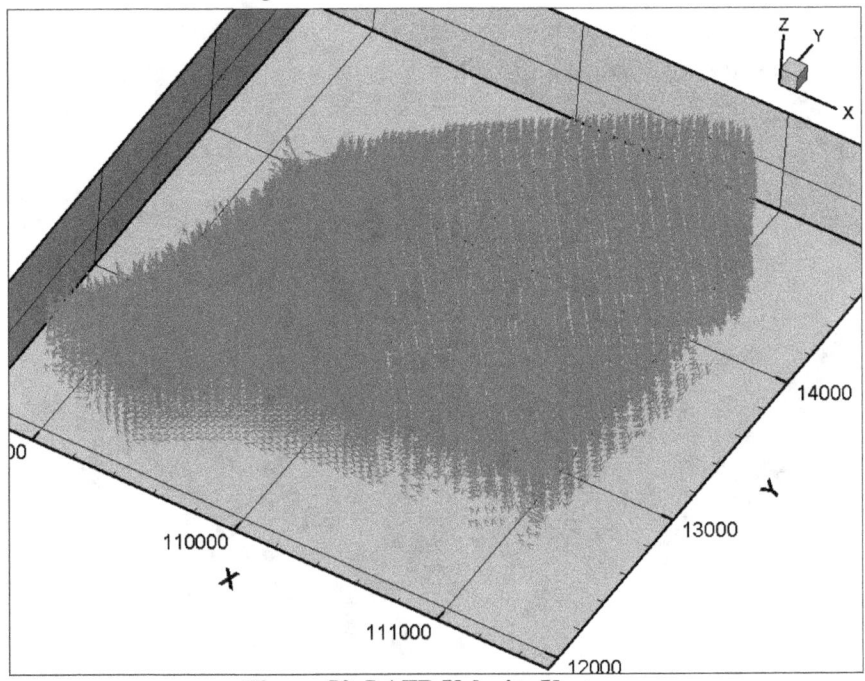

Figure 50. LAFB Velocity Vectors

These velocities may be found in file LAFB.VEC, which looks like:

```
VARIABLES="X", "Y", "Z", "U", "V", "W"
109373 12074.8 385.425 -0.22144 -0.078174 0.0270086
109401 12094.3 386.137 -0.59286 -0.35286  0.022306
109354 12102.5 385.875 -0.4435  0.055038  0.0228016
etc.
```

A corresponding layout file is provided, LAFB.LAY, to facilitate displaying this information with Tecplot™. The data file is listed at the top along with the variable names, as shown below:

```
$!VarSet |LFDSFN1| = '"LAFB.VEC"'
$!VarSet |LFDSVL1| = '"X" "Y" "Z" "U" "V" "W"'
$!SETSTYLEBASE FACTORY
$!GLOBALTHREEDVECTOR
  UVAR = 4
  VVAR = 5
  WVAR = 6
  RELATIVELENGTH = 2E-005
  HEADSIZEINFRAMEUNITS = 1
  SIZEHEADBYFRACTION = NO
```

```
$!FIELDLAYERS
  SHOWMESH = NO
  SHOWVECTOR = YES
  SHOWBOUNDARY = NO
```

The initial concentrations (two plumes) are shown in this next figure:

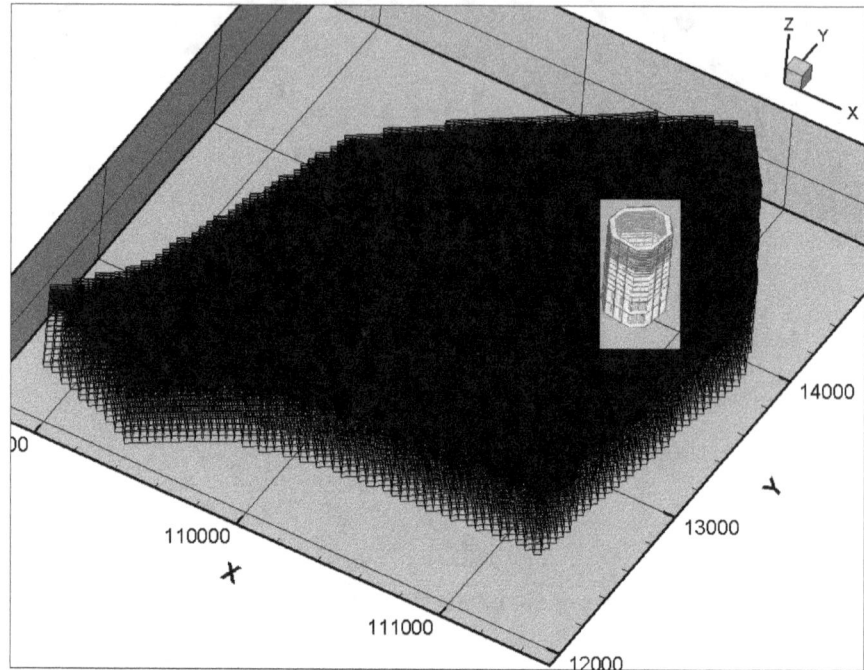

Figure 51. Initial Concentrations (log) for LAFB Example

Concentrations are written to files are named LAFB???.DAT, which can be imported by Tecplot™. The prefix is used to create the sequential output files. The same layout file (LAFB.LAY) can be used to display the concentrations. Refer to the Tecplot™ documentation for how to turn the elements on and off and control the display of concentrations, including colors. You will need to manually load the concentration file for each time step (File => Load DataFile => Replace Dataset and retain plot style).

Before we move ahead numerical solutions, we first compare this approach to the analytical solution, beginning with diffusion alone. The analytical results after 100 years is shown in this next figure:

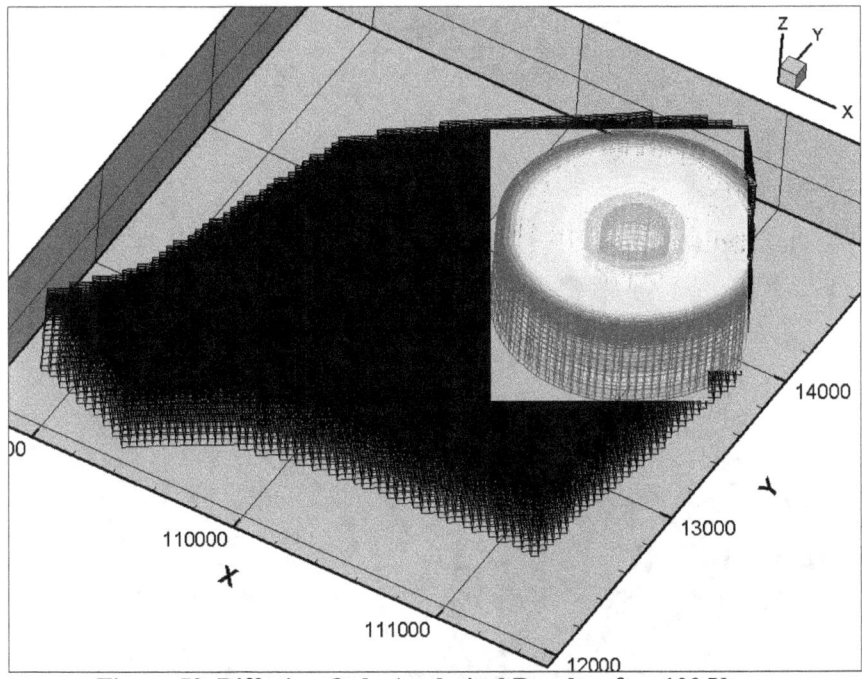

Figure 52. Diffusion Only Analytical Results after 100 Years

We use linear-least squares regression to approximate the concentration over each internal node (those having 26 neighbors) and then take the partial derivatives of the approximation:

```
for(n=0;n<Nn;n++)
  {
  if(Ln[n]<26)
    continue;
  for(i=0;i<Ln[n];i++)
    {
    l=La[26*n+i];
    C[i]=Node[l].C;
    X[i]=Node[l].X-Node[n].X;
    Y[i]=Node[l].Y-Node[n].Y;
    Z[i]=Node[l].Z-Node[n].Z;
    }
  X[i]=Y[i]=Z[i]=0.;
  C[i++]=Node[n].C;
  SecondOrderRegression(X,Y,Z,C,i,A);
  dCdX=A[1];
```

```
dCdY=A[2];
dCdZ=A[3];
d2CdX2=2.*A[4];
d2CdY2=2.*A[7];
d2CdZ2=2.*A[9];
dCdt[n]=Node[n].Dx*d2CdX2
       +Node[n].Dy*d2CdY2
       +Node[n].Dz*d2CdZ2
       -Node[n].U*dCdX
       -Node[n].V*dCdY
       -Node[n].W*dCdZ;
}
```

The numerical results are shown in this next figure:

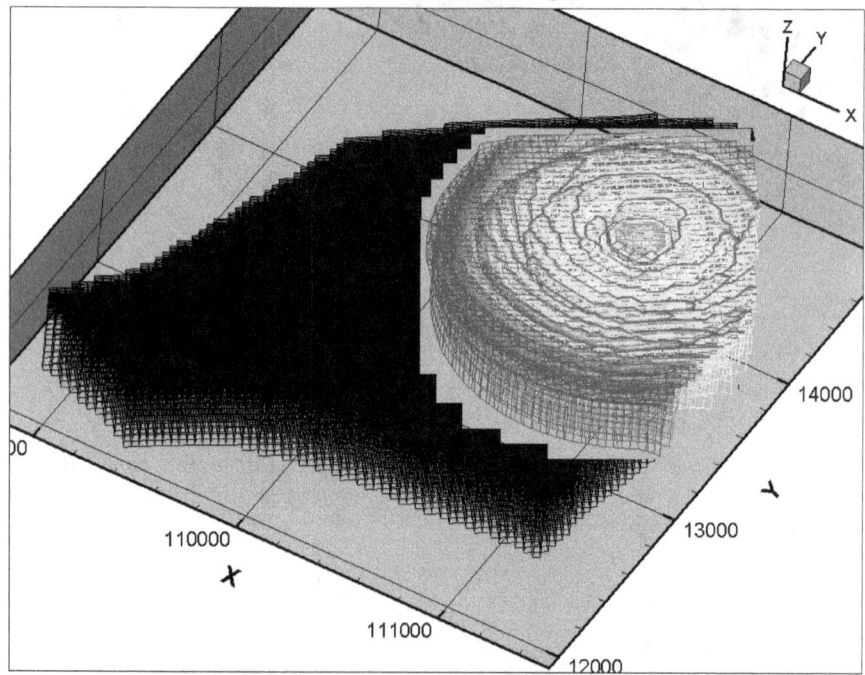

Figure 53. Diffusion Only Numerical Results after 100 Years

The colors are the same in Figures 53 and 54 (see contour levels in Tecplot™). Note that the advance of the outer blue ring corresponding to $10^{-7.3}$ ppm is the same for the analytical and numerical solutions even though there is some blotchiness toward the center of the plume.

Check and Restart

If you read through the main function in each of these example codes, you will find a check: if(Cx>Co). Unless we are adding mass or increasing toxicity over time, the maximum concentration in the domain shouldn't increase. This is an indication of a numerically unstable condition. In the previous example codes we simply wrote an error message and exited the program. In ctfra.c we restart and try again several times. Before restarting we do two things: 1) divide the time step by 2; and 2) increase the relaxation parameter.

You may be familiar with successive over relaxation (SOR) and possibly under relaxation. Over relaxation only works when solving systems of linear equations. When solving nonlinear equations, especially stiff ones, under relaxation is often necessary. The differencing scheme used here (multivariate linear regression of second order) there is a tendency to over estimate the partial derivatives. When we are simply curve-fitting and considering the target value this may not be so noticeable but when we're estimating the derivatives, it is. To quash this over estimation, we implement a simple relaxation (i.e., averaging the surrounding nodes):

```
for(r=0;r<relax;r++)
  {
  for(n=0;n<Nn;n++)
    {
    for(i=0;i<Ln[n];i++)
      {
      l=La[26*n+i];
      dCdt[n]+=dCdt[l];
      }
    dCdt[n]/=Ln[n]+1;
    }
  }
```

Launching the program (ctfra.c) for this example produces:

```
ctfra lafb
3D Transient Diffusion + Dispersion + Advection
reading FRAC3D input and output files
prefix: lafb
node file: LAFB.NDE
  nodes: 99372
  108830≤X≤111418
  12051≤Y≤14764
  359.5≤Z≤701.8
element file: LAFB.ELM
  elements: 91575
  rotation: 36°
property file: LAFB.PRP
  73.05≤Dx≤73.05
  73.05≤Dy≤73.05
  7.305≤Dz≤7.305
  0.05≤F≤0.05
velocity file: LAFB.VEP
  -186.66≤U≤262.608
  -143.844≤V≤29.583
```

```
   -25.2326≤W≤30.5568
velocities: LAFB.VEC
determining time step
   ΔX=47, ΔY=47, ΔZ=5.4
   ΔX²/Dx/2=15.1198, ΔY²/Dy/2=15.1198, ΔZ²/Dz/2=1.99589
   ΔX/U/6=0.029829, ΔY/V/6=0.0544571, ΔZ/W/6=0.0294533
   Δt=0.0115643
neighbors: 2442220
   internal nodes: 84144
   boundary nodes: 15228
concentration file: LAFB.ICN
   C≤300
0 years C≤300 LAFB000.DAT
1 years C≤300 LAFB001.DAT
```

The concentrations after 100 years with diffusion/dispersion:

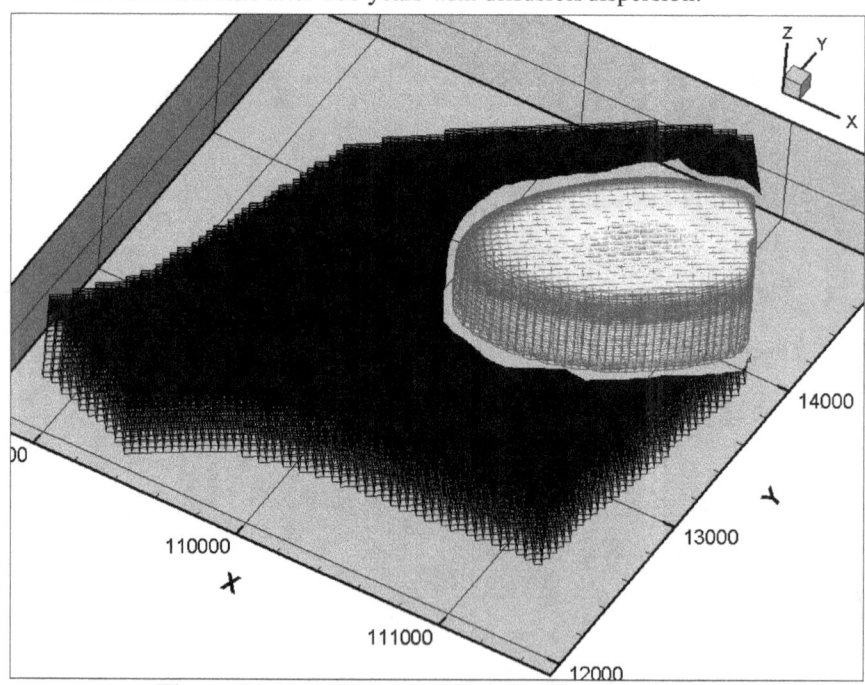

Figure 54. Full Numerical Results after 100 Years

PORTS Example

This next FRAC3D based example can be found in the online archive in folder examples\FRAC3D\PORTS. The elements are shown in this first figure:

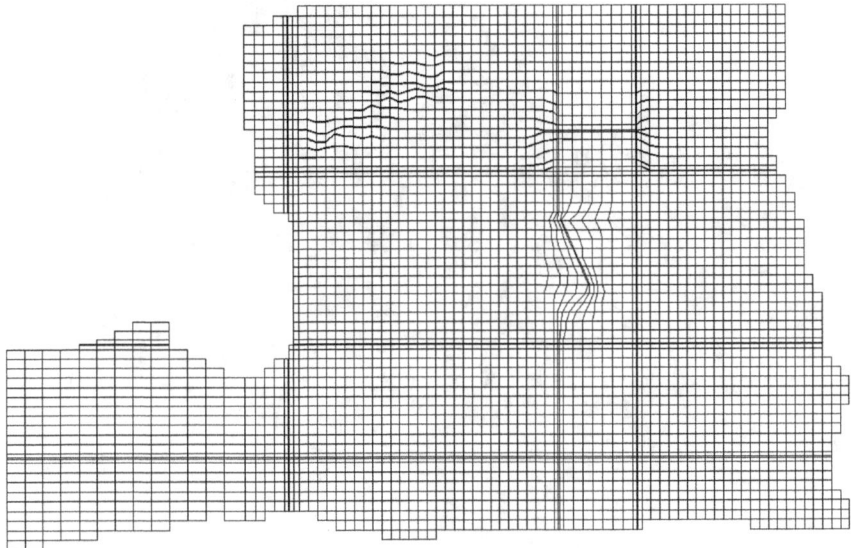

Figure 55. Elements for PORTS Example

The elements are of non-uniform size and a few are rotated but this will not greatly impact our calculations here. These adjustments were made to the original model to investigate the impact of fractures using particle tracking, as karst formations are common at the site. Further drilling and investigations were conducted to better characterize the ground properties. Many model revisions were created before settling on one that was used to guide the remediation effort.

The velocity vectors are shown in this next figure:

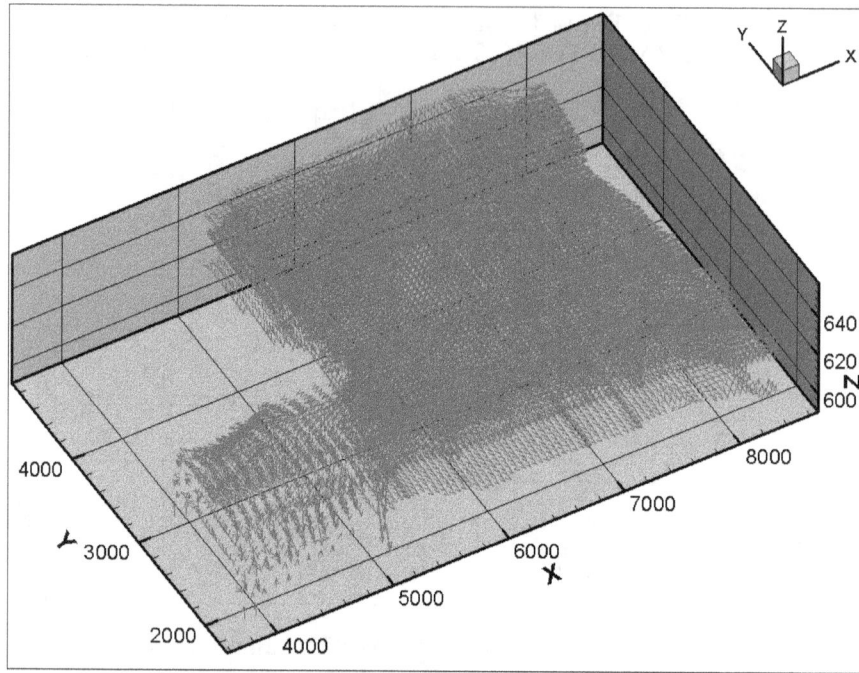

Figure 56. PORTS Example Velocity Vectors

Velocities varied throughout the domain for this particular site. Notice the changes at the edge on the southwest side in particular. This odd shape arose from construction in the area, including buildings and roads. All of these factors were considered in building the model and designing the remediation strategy.

While there were several plumes at this site and several contaminants involved, we will consider a single plume for our example here. We could add a term for each contaminant and calculate them simultaneously or run the model several times, once for each contaminant species. The initial concentrations are shown in this next figure:

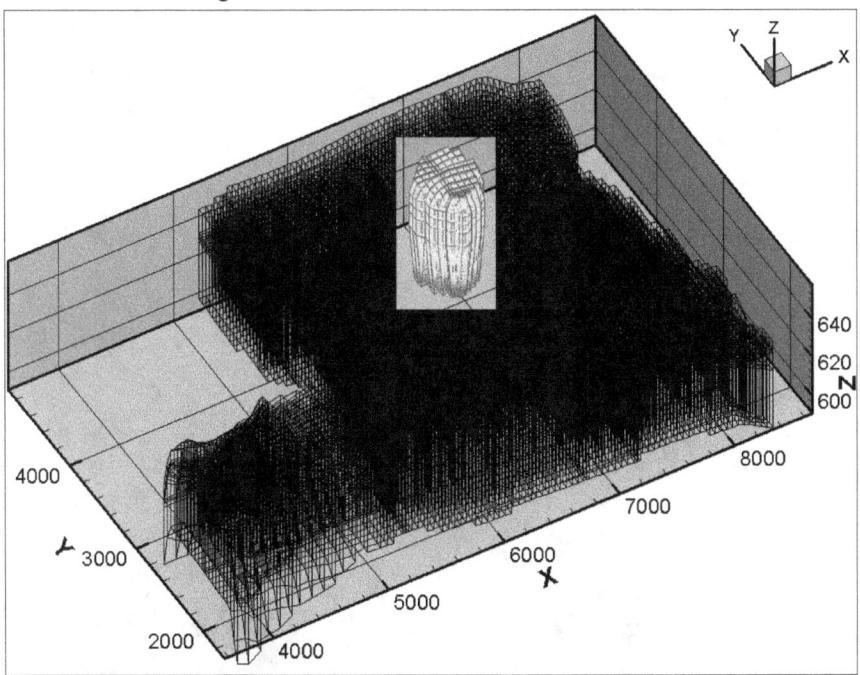

Figure 57. Initial Concentrations (log) for PORTS Example

Launching the program produces the following output:

```
ctfra ports
3D Transient Diffusion + Dispersion + Advection
reading FRAC3D input and output files
prefix: ports
node file: PORTS.NDE
  nodes: 26844
  3750≤X≤8460
  1795≤Y≤4792
  575.4≤Z≤665
element file: PORTS.ELM
  elements: 21500
  rotation: 0°
property file: PORTS.PRP
  365.25≤Dx≤365.25
  365.25≤Dy≤365.25
  36.525≤Dz≤36.525
```

```
   0.1≤F≤0.1
velocity file: PORTS.VEP
   -305.66≤U≤351.18
   -768.92≤V≤185.26
   -23.56≤W≤7.84
velocities: PORTS.VEC
determining time step
   ΔX=50, ΔY=53, ΔZ=4.6
   ΔX²/Dx/2=3.42231, ΔY²/Dy/2=3.84531, ΔZ²/Dz/2=0.289665
   ΔX/U/6=0.0237295, ΔY/V/6=0.011488, ΔZ/W/6=0.032541
   Δt=0.00610038
neighbors: 600676
   internal nodes: 16512
   boundary nodes: 10332
concentration file: PORTS.ICN
   C≤300
0 years C≤300 PORTS000.DAT
1.000 years C≤284 PORTS001.DAT
36.017 years C≤3 PORTS036.DAT
```

The concentrations after 10 years are shown in this next figure:

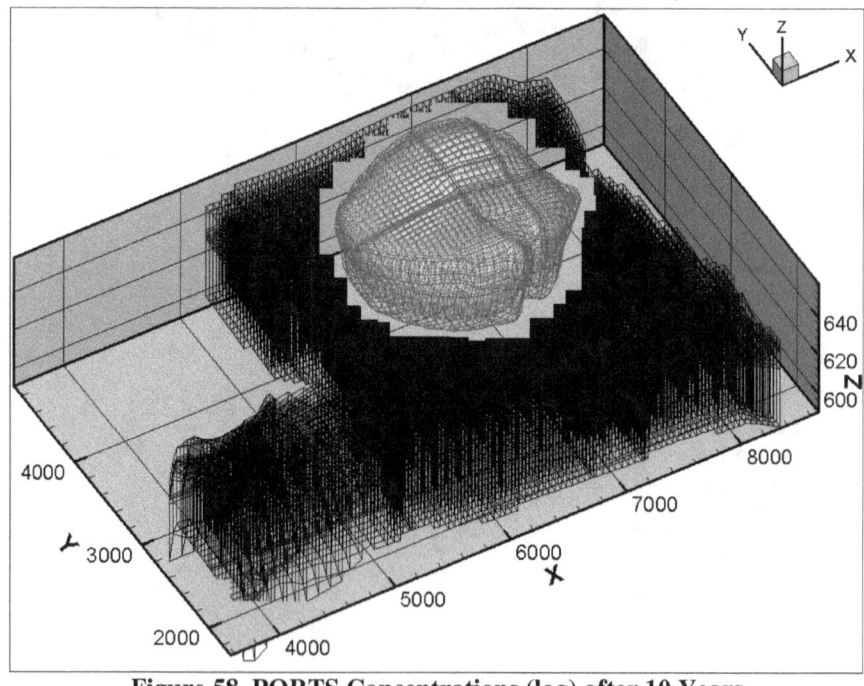

Figure 58. PORTS Concentrations (log) after 10 Years

The interior of the plume is obscured by the top. Tecplot™ has a feature (menu => style => value blanking) that can turn off drawing the top of the plume, thus exposing the interior. The selection and result are shown in this next figure:

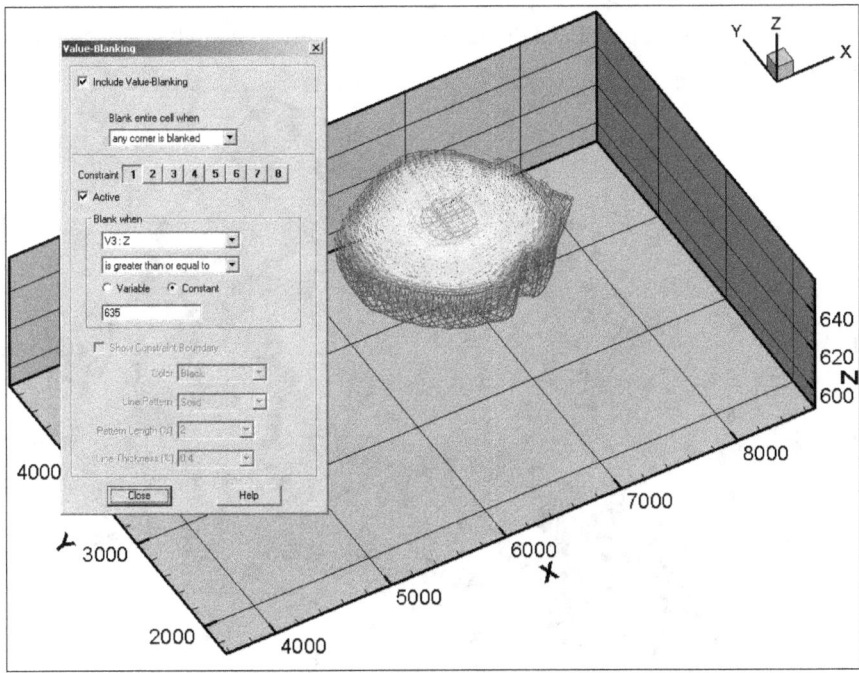

Figure 59. Concentrations with Value Blanking

Value blanking can also be used with a macro to slice through the plume in any of the three directions, exporting the graphic each time to sequential files. The following is an example of such a Tecplot™ macro:

```
#!MC 800
$!REDRAWALL
$!EXPORTSETUP EXPORTFORMAT = RASTERMETAFILE
$!EXPORTSETUP EXPORTFNAME = 'plume.rm'
$!FIELD [1]   IJKMODE{PLANES = K}
$!FIELD [1]   IJKMODE{KRANGE{MIN = 1}}
$!FIELD [1]   IJKMODE{KRANGE{MAX = 1}}
$!REDRAWALL
$!EXPORT
  APPEND = NO
$!FIELD [1]   IJKMODE{KRANGE{MIN = 2}}
$!FIELD [1]   IJKMODE{KRANGE{MAX = 2}}
$!REDRAWALL
$!EXPORT
  APPEND = YES
```

```
$!FIELD [1]   IJKMODE{KRANGE{MIN = 3}}
$!FIELD [1]   IJKMODE{KRANGE{MAX = 3}}
$!REDRAW
$!EXPORT
  APPEND = YES
```

The plume has spread and dissipated considerably so that after 36 years, this is what's left:

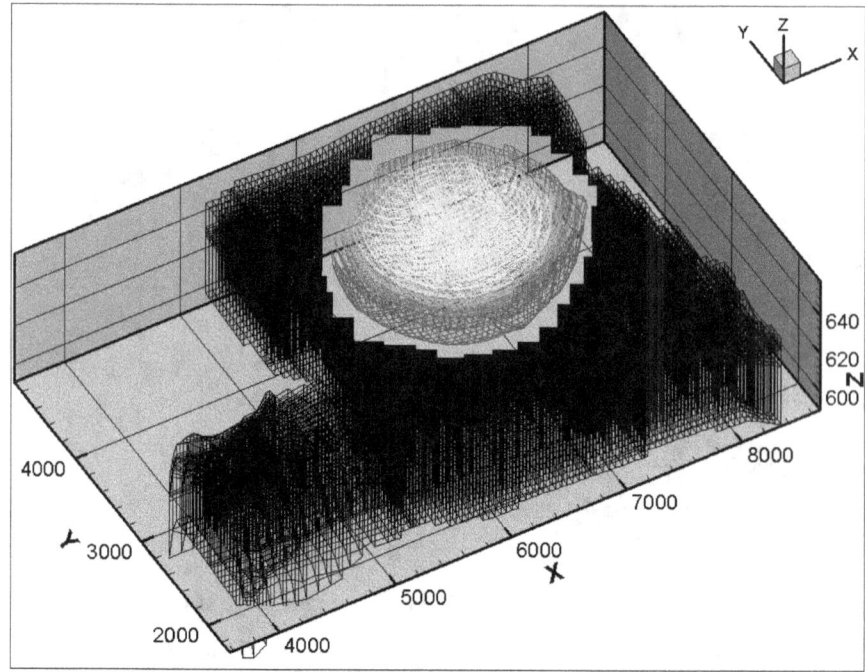

Figure 60. PORTS Concentrations (log) after 36 Years

ORNL Example

The elements for this next example are shown in the following figure:

Figure 61. ORNL Example Elements

The velocity vectors are shown below:

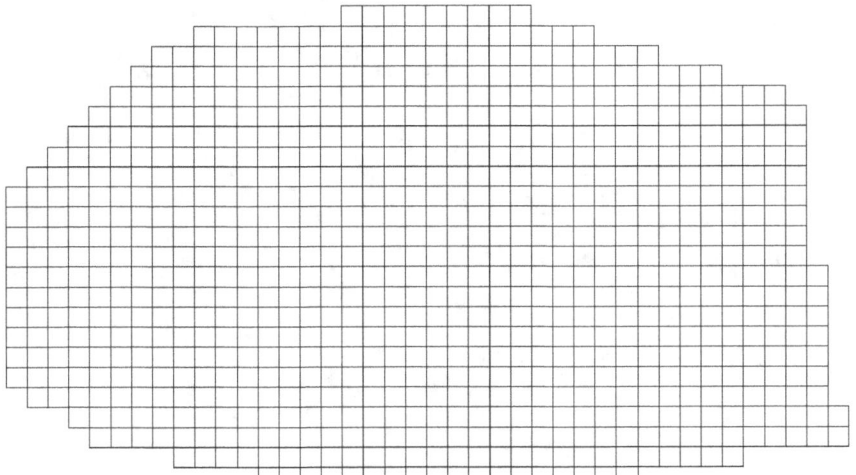

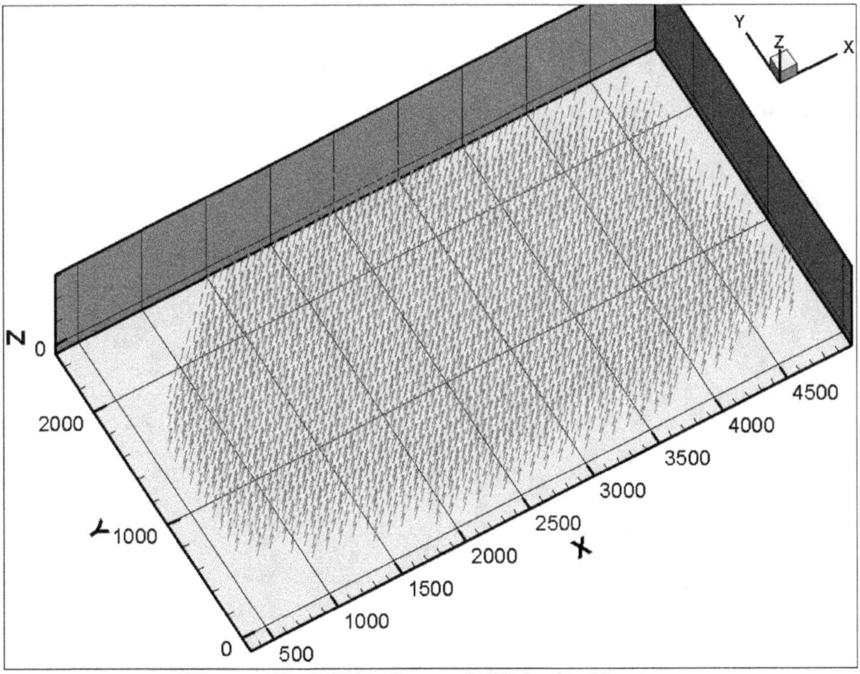

Figure 62. ORNL Example Velocity Vectors

There were also several plumes at this site but we consider only one here:

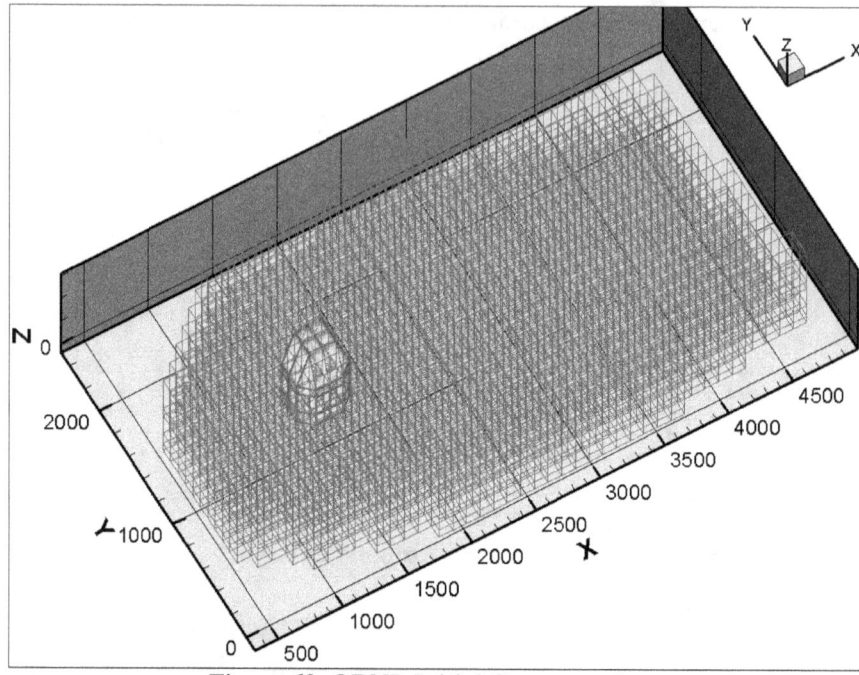

Figure 63. ORNL Initial Concentrations

Launching the program produces the following output:
```
ctfra ornl
3D Transient Diffusion + Dispersion + Advection
reading FRAC3D input and output files
prefix: ornl
node file: ORNL.NDE
  nodes: 5166
  532≤X≤4787
  0≤Y≤2400
  -23.7≤Z≤294.3
element file: ORNL.ELM
  elements: 3980
  rotation: 0°
property file: ORNL.PRP
  365.25≤Dx≤365.25
  365.25≤Dy≤365.25
  36.525≤Dz≤36.525
  0.1≤F≤0.1
velocity file: ORNL.VEP
  500≤U≤500
  100≤V≤100
```

```
  10≤W≤10
velocities: ORNL.VEC
determining time step
  ΔX=106, ΔY=100, ΔZ=125.6
  ΔX²/Dx/2=15.3812, ΔY²/Dy/2=13.6893, ΔZ²/Dz/2=215.953
  ΔX/U/6=0.0353333, ΔY/V/6=0.166667, ΔZ/W/6=2.09333
  Δt=0.028635
neighbors: 112546
  internal nodes: 2932
  boundary nodes: 2234
concentration file: ORNL.ICN
  C≤300
0 years C≤300 ORNL000.DAT
1.002 years C≤164 ORNL001.DAT
```

This is a small, rapidly spreading plume. After 1 year the results are:

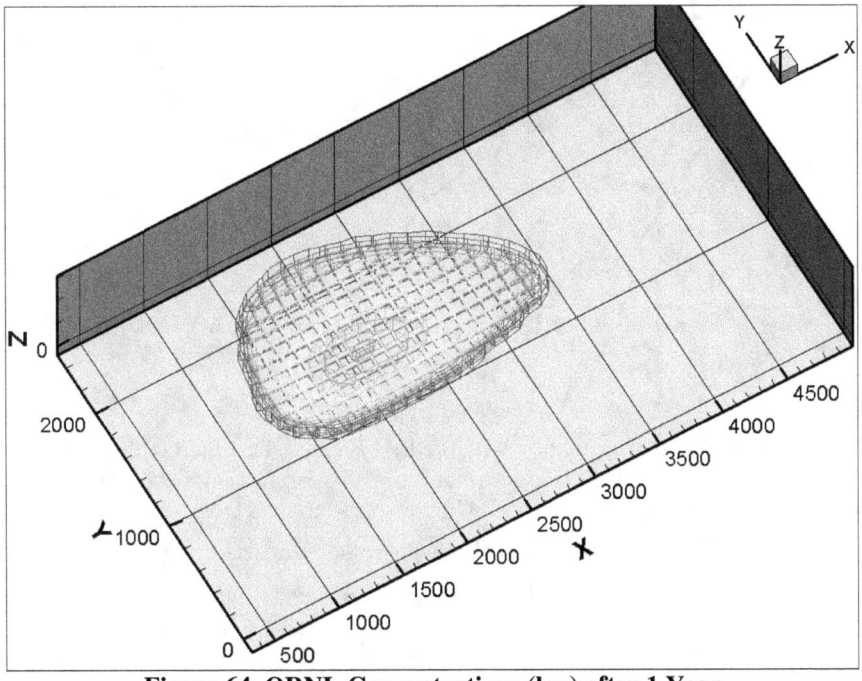

Figure 64. ORNL Concentrations (log) after 1 Year

The plume quickly dissipates and exits the site (we lost this one). After 6 years it looks like this:

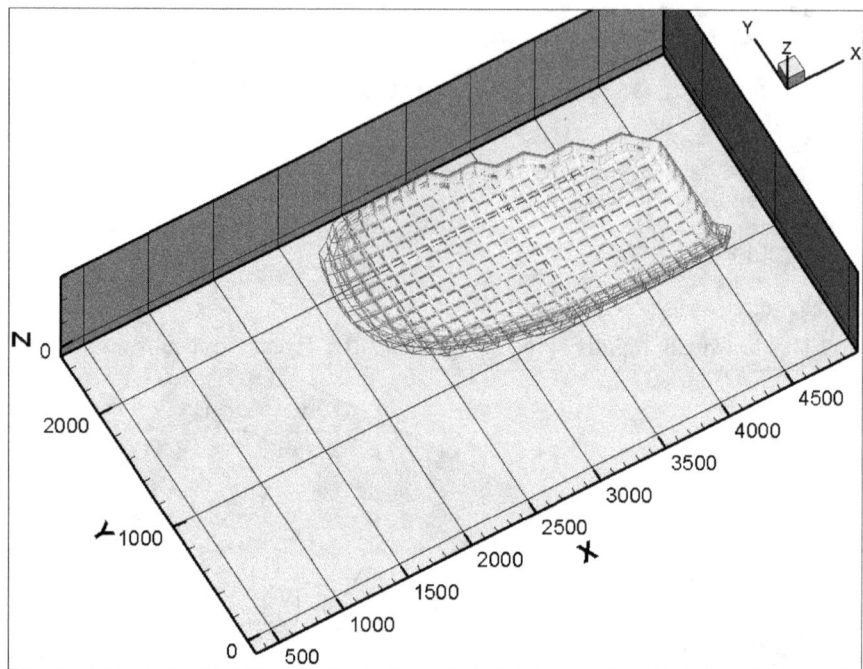

Figure 65. ORNL Concentrations (log) after 6 Years

Rotated Elements

We introduced grid rotation with the LAFB example, which in that case was 36°. Each element has a different rotation in this next example. As before, we calculate the finite differences with respect to the two planar element coordinates $\bar{P}$ and $\bar{Q}$, then multiply by the rotation matrix:

$$\begin{bmatrix} \cos\theta & -\sin\theta \\ \sin\theta & \cos\theta \end{bmatrix} \quad (6.1)$$

We use the rotation angles with the nodal equations and so we transfer the angle from each element to the first node. The rotated difference equations become:

```
for(n=0;n<Nn;n++)
  {
  if(Ln[n]<26) /* only consider internal nodes */
    continue;
```

```
    dP=hypot(Node[Ie[n]].Y-Node[Iw[n]].Y,Node[Ie[n]].X-
Node[Iw[n]].X)/2.;
    dQ=hypot(Node[In[n]].Y-Node[Is[n]].Y,Node[In[n]].X-
Node[Is[n]].X)/2.;
    dZ=(Node[Iu[n]].Z-Node[Il[n]].Z)/2.;
    Cc=Node[n].C;
    Ce=Node[Ie[n]].C;
    Cn=Node[In[n]].C;
    Cs=Node[Is[n]].C;
    Cw=Node[Iw[n]].C;
    Cu=Node[Iu[n]].C;
    Cl=Node[Il[n]].C;
    Ca=(Cw+Cc)/2.;
    Cb=(Ce+Cc)/2.;
    Cd=(Cn+Cc)/2.;
    Cf=(Cs+Cc)/2.;
    Cg=(Cu+Cc)/2.;
    Ch=(Cl+Cc)/2.;
    dCdPa=(Cc-Cw)/dP;
    dCdPb=(Ce-Cc)/dP;
    dCdQd=(Cn-Cc)/dQ;
    dCdQf=(Cc-Cs)/dQ;
    dCdZg=(Cu-Cc)/dZ;
    dCdZh=(Cc-Cl)/dZ;
    d2CdP2=(dCdPb-dCdPa)/dP;
    d2CdQ2=(dCdQd-dCdQf)/dQ;
    d2CdZ2=(dCdZg-dCdZh)/dZ;
    Dx=Node[n].Dx;
    Dy=Node[n].Dy;
    Dz=Node[n].Dz;
    U=Node[n].U;
    V=Node[n].V;
    W=Node[n].W;
    if(U>0.)
        UdCdX=U*(cos*(Cc-Cw)/dP-sin*(Cc-Cs)/dQ);
    else
        UdCdX=U*(cos*(Ce-Cc)/dP-sin*(Cn-Cc)/dQ);
    if(V>0.)
        VdCdY=V*(sin*(Cc-Cw)/dP+cos*(Cc-Cs)/dQ);
    else
        VdCdY=V*(sin*(Ce-Cc)/dP+cos*(Cn-Cc)/dQ);
    if(W>0.)
        WdCdZ=W*(Cc-Cl)/dZ;
    else
        WdCdZ=W*(Cu-Cc)/dZ;
    dCdt[n]=Dx*(co*d2CdP2-si*d2CdQ2)
           +Dy*(si*d2CdP2+co*d2CdQ2)
           +Dz*d2CdZ2
           -UdCdX-VdCdY-WdCdZ;
```

The Z direction is not rotated so no modification is required. This rotation and also the relaxation and restart procedure are two examples of how you might need to modify a model to handle numerical issues as they arise. The elements for this last FRAC3D-based example are all slightly rotated:

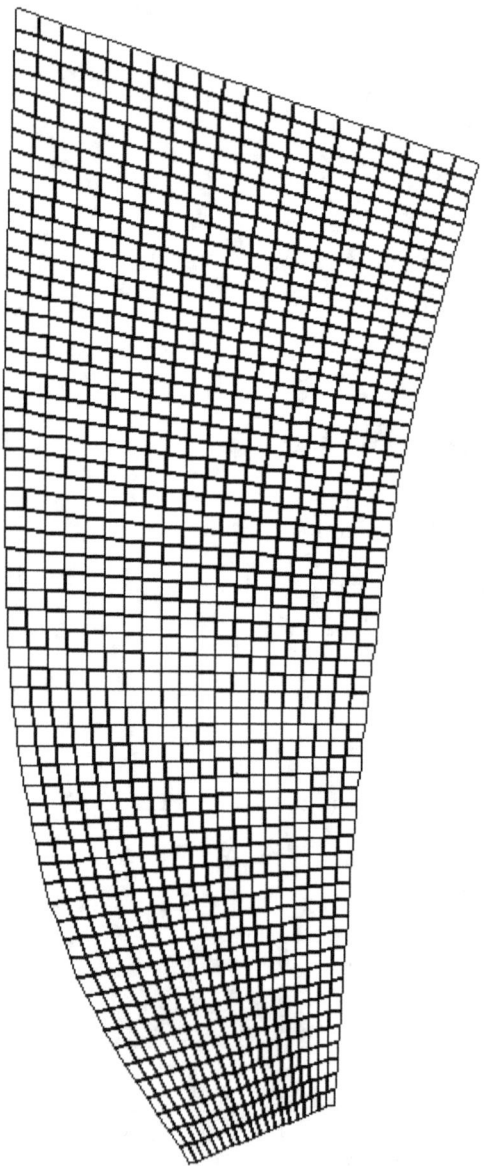

Figure 66. CAFB Example Elements

The velocity vectors for this example are shown in the following figure:

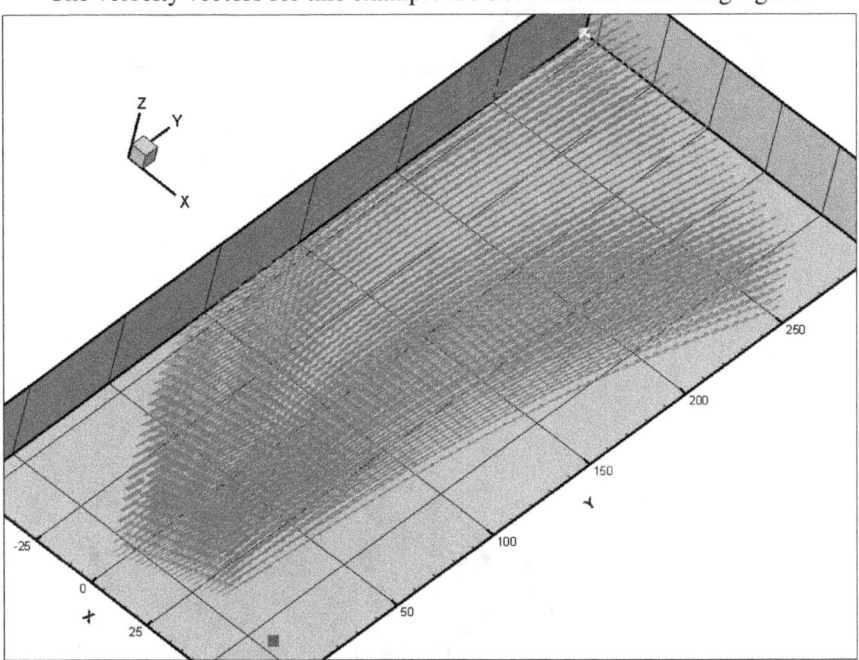

Figure 67. CAFB Example Velocity Vectors

Launching the program produces the following output:

```
ctfra cafb
3D Transient Diffusion + Dispersion + Advection
reading FRAC3D input and output files
prefix: cafb
node file: CAFB.NDE
  nodes: 10248
  -52.2≤X≤71.5
  -11.6≤Y≤286.9
  0≤Z≤35
element file: CAFB.ELM
  elements: 8400
  rotation: 124°
property file: CAFB.PRP
  73.05≤Dx≤73.05
  73.05≤Dy≤73.05
  3.6525≤Dz≤3.6525
  0.1≤F≤0.1
velocity file: CAFB.VEP
  20≤U≤20
  200≤V≤200
  2≤W≤2
```

```
determining time step
  ΔX=7.1, ΔY=5.9, ΔZ=5
  ΔX²/Dx/2=0.345038, ΔY²/Dy/2=0.238261, ΔZ²/Dz/2=3.42231
  ΔX/U/6=0.0591667, ΔY/V/6=0.00491667, ΔZ/W/6=0.416667
  Δt=0.00434634
neighbors: 232654
  internal nodes: 6726
  boundary nodes: 3522
concentration file: CAFB.ICN
  C≤300
0 years C≤300 CAFB000.DAT
1 years C≤171 CAFB001.DAT
```

The initial concentrations are shown below:

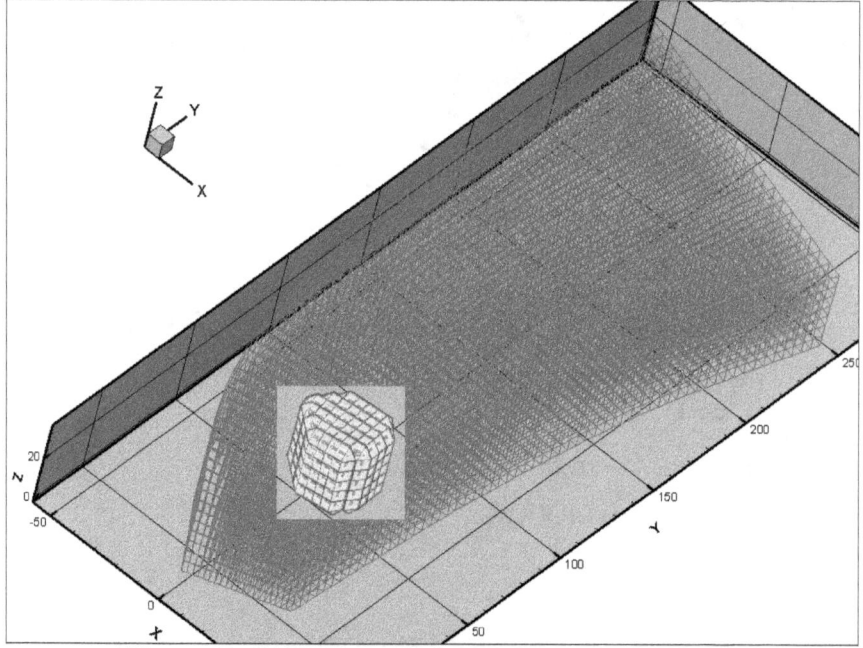

Figure 68. CAFB Example Initial Concentrations (log)

This is a small domain with a rapidly dissipating and sweeping plume. After just 5 years it looks like this:

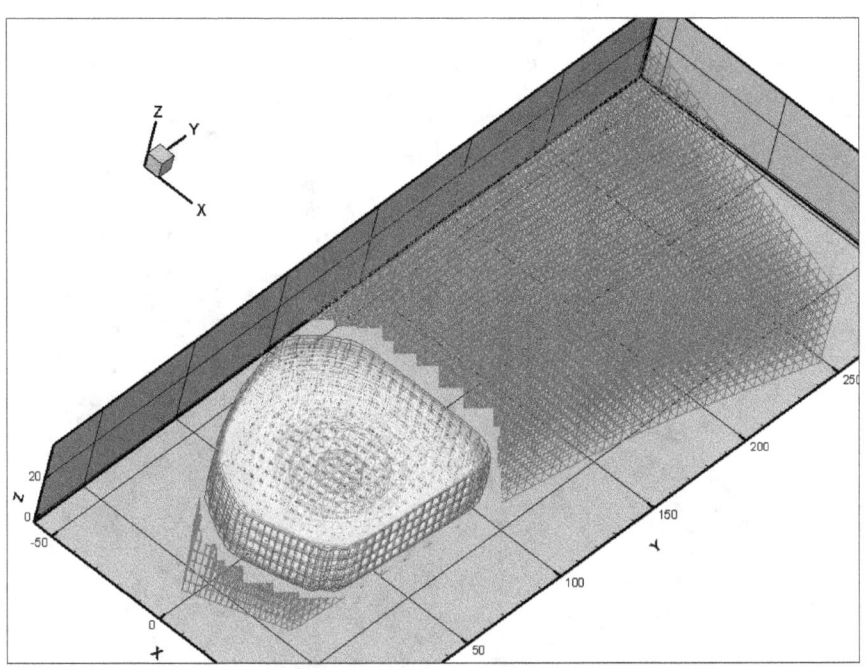

Figure 69. CAFB Concentrations (log) after 5 Years

The plume continues to sweep toward the northern boundary. Extremely small time steps were required to achieve stability, which is why the contours are so smooth. After 25 years the plume looks like this:

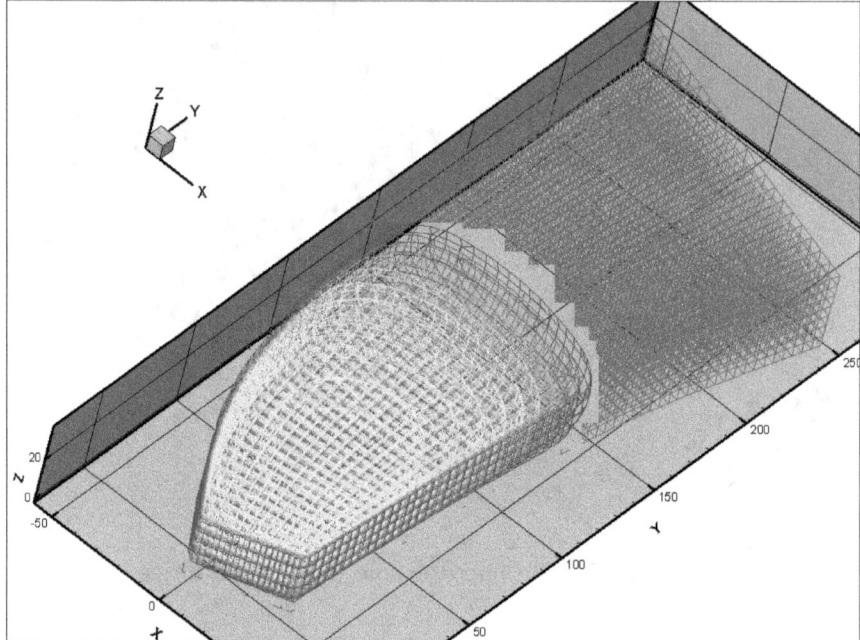

Figure 70. CAFB Concentrations (log) after 25 Years

Chapter 7. Pump-and-Treat

One of the most effective remediation strategies for removing contaminants from the ground is pump-and-treat. Groundwater is pumped from one or more wells and fed through a series of drums containing "magic dust" (i.e., some reacting and/or absorbing media that will remove the contaminant and can be disposed of properly or incinerated). In order for this approach to be effective, the pattern of movement of the contaminant through the ground must be accurately understood, which requires not only calculating but also verifying the flow. Once this has been sufficiently established, you don't want to change anything, so simply pumping water out of the ground might alter the flow pattern and decrease effectiveness. That is why the contaminated water is most often treated and then injected back into the ground nearby, as shown in this figure:

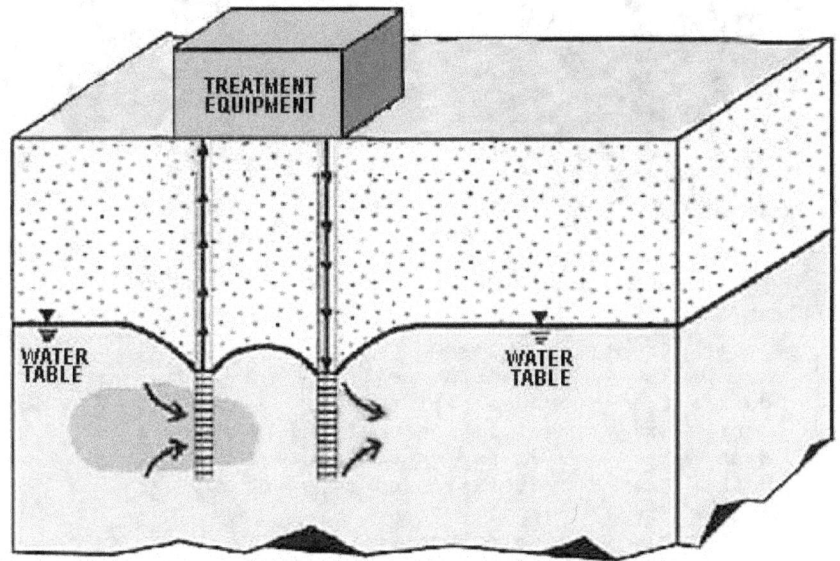

Figure 71. Pump-and-Treat Concept

The equipment to accomplish this varies but most often includes one or more pumps and one or more barrels of magic dust. One project I worked on required three sequential passes through the extraction material to adequately reduce the concentration. The first barrel became fouled most quickly so that each month the three barrels were rotated, discarding the first one and adding a fresh one to the end of the process. Four sets of barrels were used so that twelve were in use at all times with four waiting deployment. The pumps were run by dozens of automotive batteries, which were recharged weekly. All of the equipment was sunk in a concrete vault in the ground with a locked steel door to prevent tampering and theft, which occurred on a previous job. The equipment looked something like this:

Figure 72. Typical Pump-and-Treat Equipment

Implementing this in the model is simple. A switch is added to the node structure:

```
typedef struct{
  int well;    /* extraction well */
  double C;    /* concentration */
  double Dx;   /* X diffusion coefficient */
  double Dy;   /* Y diffusion coefficient */
  double Dz;   /* Z diffusion coefficient */
  double F;    /* porosity */
  double U;    /* X velocity */
  double V;    /* Y velocity */
  double W;    /* Z velocity */
  double X;    /* location */
  double Y;    /* location */
  double Z;    /* location */
  double co;   /* cos(rotation) */
  double si;   /* sin(rotation) */
  }NODE;
NODE*Node;
```

Nodes are flagged by their position:
```
void AddWell(double X,double Y,double R)
  {
  int i,n;
```

```
Nw++;
for(i=n=0;n<Nn;n++)
   if(hypot(Node[n].X-X,Node[n].Y-Y)<=R)
      {
      Node[n].well=Nw;
      i++;
      }
printf("%i nodes in well %i\n",i,Nw);
}
```

Wells are added as necessary:

```
AddWell(0.,100.,5.);
```

As the solution progresses, the concentration at all flagged nodes is set to zero:

```
for(n=0;n<Nn;n++)
   if(Node[n].well)
      Node[n].C=0.;
   else if(Ln[n]>=26)
      Node[n].C=dCdt[n]/(Ln[n]+1);
```

This is what a well looks like in the CAFB example:

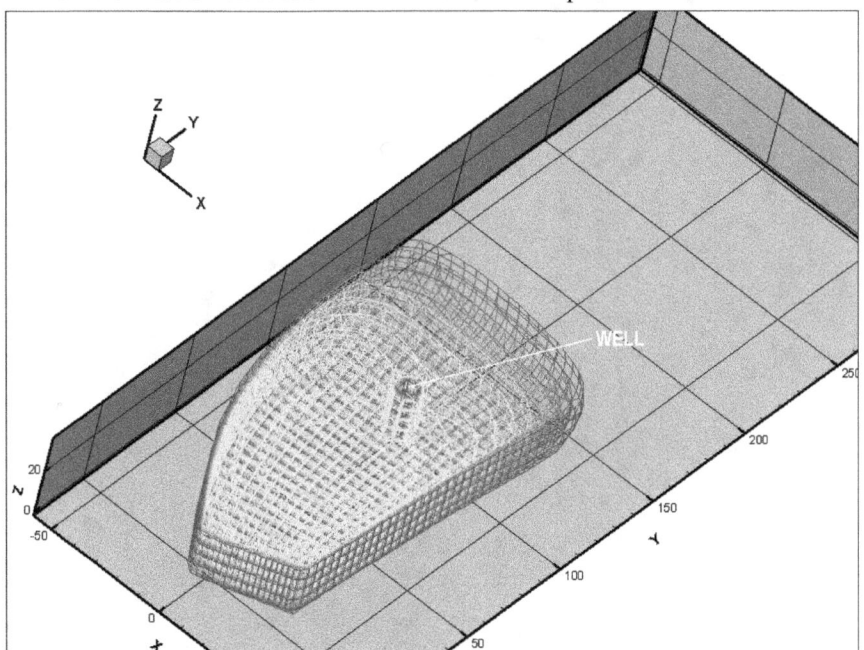

Figure 73. Well Added to CAFB Example

Chapter 8. Reaction and Decay

Chemical reactions and radioactive decay occur in some contaminants. Both can usually be approximated by the standard exponential decay formula for concentration over time:

$$C(t) = C_0 e^{-\frac{t}{\tau}} \qquad (8.1)$$

The rate of change is easily calculated:

$$\frac{dC}{dt} = \frac{-C_0}{\tau} e^{-\frac{t}{\tau}} = \frac{-C}{\tau} \qquad (8.2)$$

Equation 8.1 is shown in the following figure:

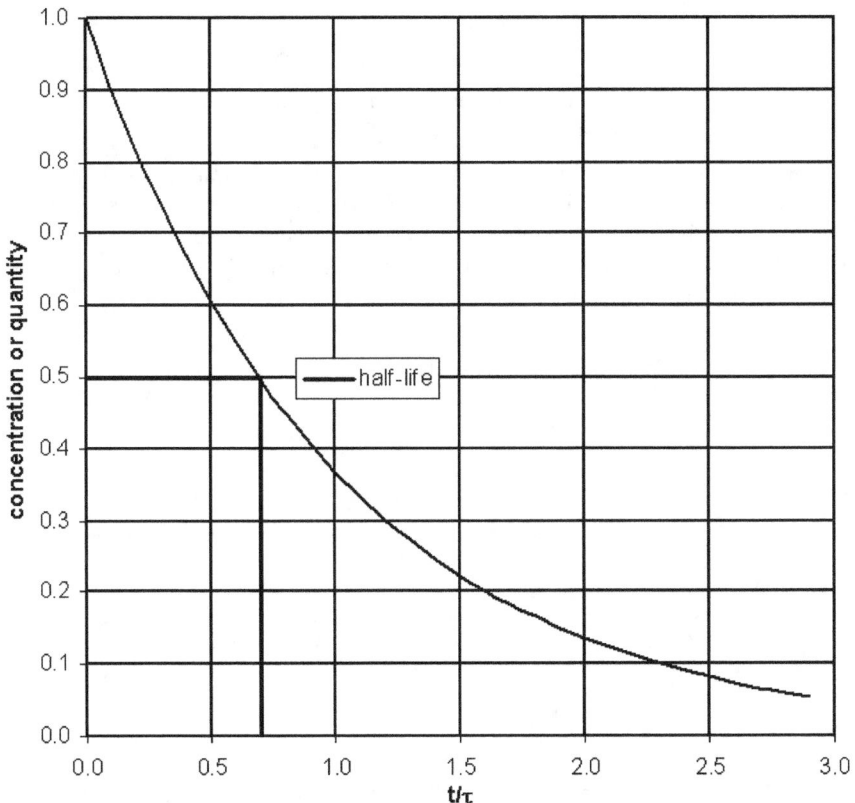

Figure 74. Exponential Decay

The "half-life" is the time required for the concentration (or quantity) to reach one half of the initial value. This happens at $\exp(-t/\tau)=0.5$ or $t/\tau=\ln(2)$. This is easily implemented in function AdvanceSolution() any one of the codes;

for example in ctmod.c we add the last term to the concentration update for each node:

```
decay=exp(-(t+dt)/tau)/exp(-t/tau);
for(z=1;z<Nlay;z++)
  {
  for(y=1;y<Nrow;y++)
    {
    for(x=1;x<Ncol;x++)
      {
      ic=node(x,y,z);
      if(!Node[ic].active)
         continue;
      Node[ic].C+=dt*dCdt[ic];
      Node[ic].C*=decay;
      }
    }
  }
```

While we might implement decay by using Equation 8.2:

```
Node[ic].C+=dt*dCdt[ic]-Node[ic].C/tau;
```

this may lead to overshoot and sometimes negative concentrations. This problem is easily avoided by simply multiplying by the ratio of exponentials as shown above. Using the WINE example and $\tau=5$ years, the corresponding half-life is 3.47 years. Concentrations with and without decay after 25 years are shown on the following page for comparison:

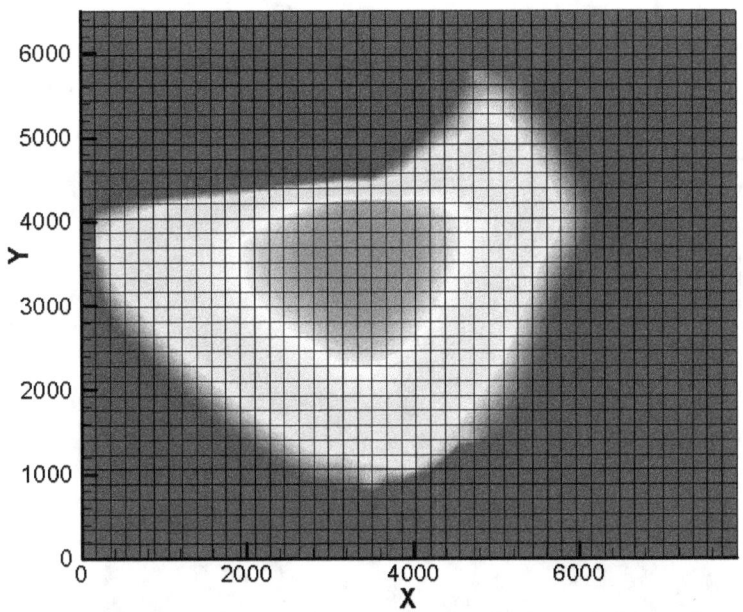

Figure 75. WINE Concentrations (log) after 25 Years without Decay

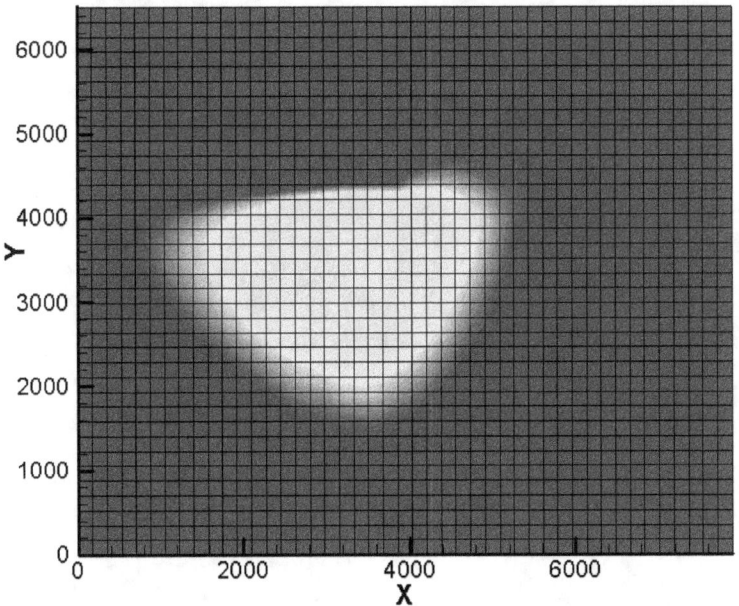

Figure 76. WINE Concentrations (log) after 25 Years with Decay

Chapter 9. AT123D Analytical Solution

AT123D is a numerical implementation of an analytical solution to the governing partial differential equation expressing the conservation of mass for the contaminant, Equation 2.8, in 1D, 2D, or 3D. The original FORTRAN code was developed by Dr. G. T. Yeh of the Oak Ridge National Laboratory.[6] A Web search will turn up documents, examples, models, and more. This useful code has been the basis of many successful remediation projects, as the literature reflects. AT123D was also used to validate my particle tracker, PTRAX.[7] It can also be used to validate other numerical methods, such as the ones accompanying this text, which can be found in the online archive in the example folders.

Because AT123D is an analytical solution, it is somewhat limited in what it can do because of the simple boundary conditions and domain shapes. An analytical model also cannot handle discontinuous properties, such as occur across soil type boundaries and rock formations, let alone karst or fractures. Still, it is a very useful tool. At one time the AT123D FORTRAN code could be downloaded free of charge from the ORNL web site, as it was developed by the Department of Energy. AT123D has been incorporated into and is available with at least one commercial package (SEVIEW).

Twenty-three years ago I translated the original FORTRAN into C and significantly optimized the code. I also replaced the crude (and slow) error function approximation with a better one from the standard reference on such.[8] I replaced the hardwired arrays with allocatable ones and modified the output to create files that can be read by TP2 or Tecplot™. I haven't made this optimized code available on the Web, as I have not been released to do so. Should you need it and can get permission from DoE to use it, I'd be glad to send it to you.

---

[6] G. T., Yeh, "AT123D: Analytical One-, Two-, and Three-Dimensional Simulation of Waste Transport in the Aquifer System," Environmental Sciences Division Publication No. 1439, Report. ORNL-5602, 1981.

[7] Benton, D. J., Young, S. C., and Williams, N. J., "Description and Verification of PTRAX - A Random Walk Model for Predicting Groundwater Solute Transport," U.S. Department of Energy, Report MMESD 8.13-005, 1995.

[8] Abramowitz, M. and I. A. Stegun, *Handbook of Mathematical Functions*, first published by the National Bureau of Standards as Technical Monograph No. 55. This valuable reference may be obtained free online as a PDF from several different web sites.

We can use AT123D to illustrate several processes. This first pair of figures shows a simple spreading, advecting plume.

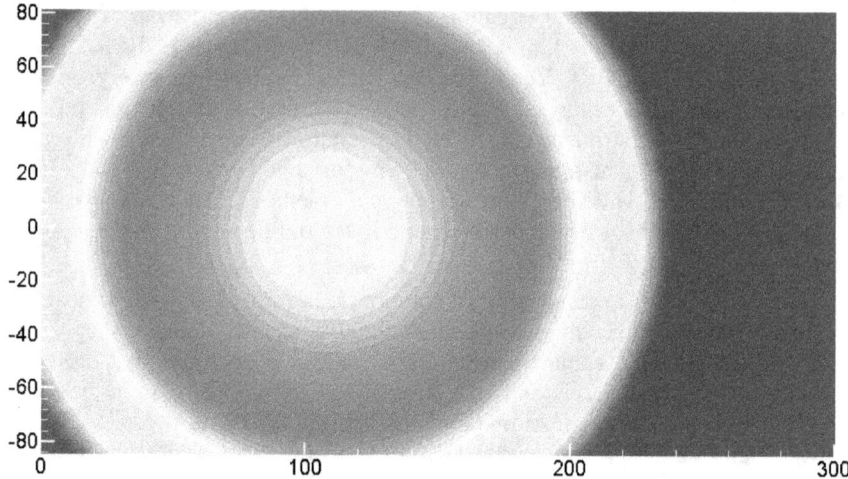

Figure 77. Diffusion/Advecting Plume after 15 Years

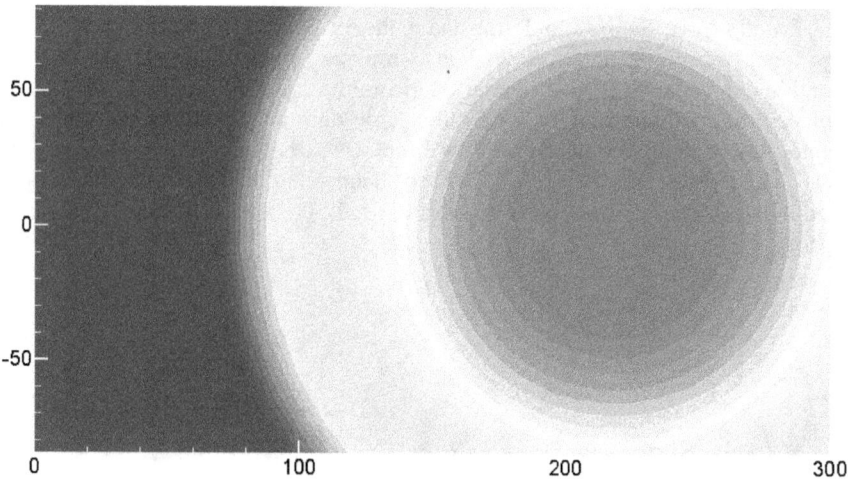

Figure 78. Diffusion/Advecting Plume after 30 Years

If we add decay and hold everything else constant, we get:

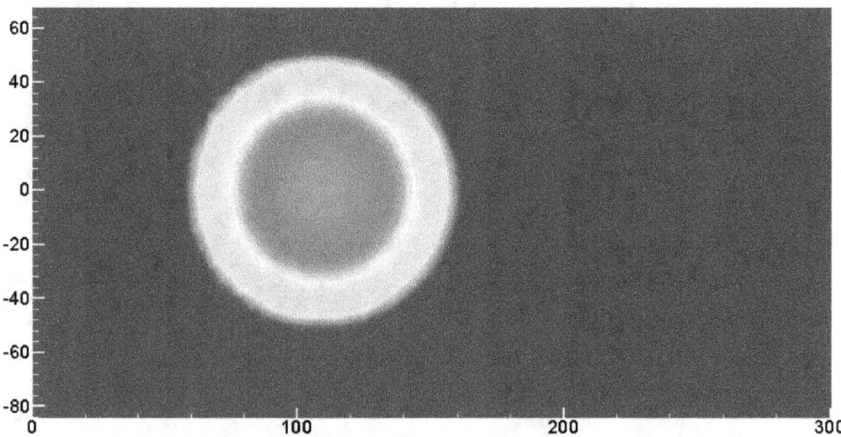

Figure 79. Decaying Plume after 15 Years

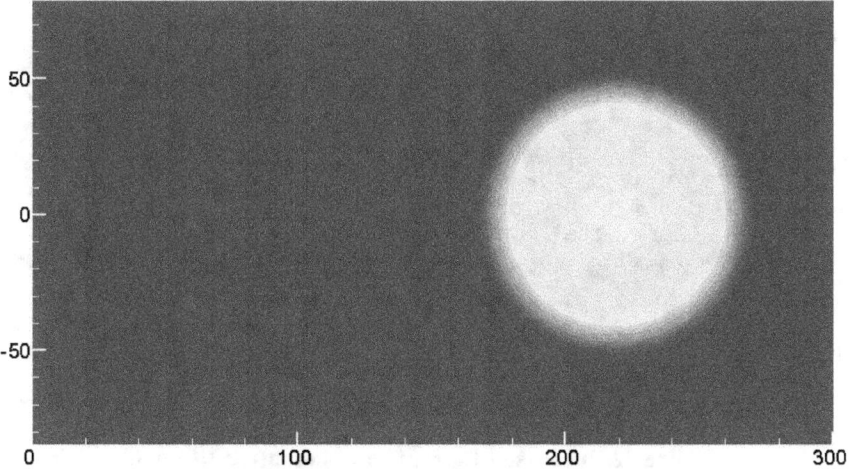

Figure 80. Decaying Plume after 30 Years

If we increase the lateral diffusion, we get:

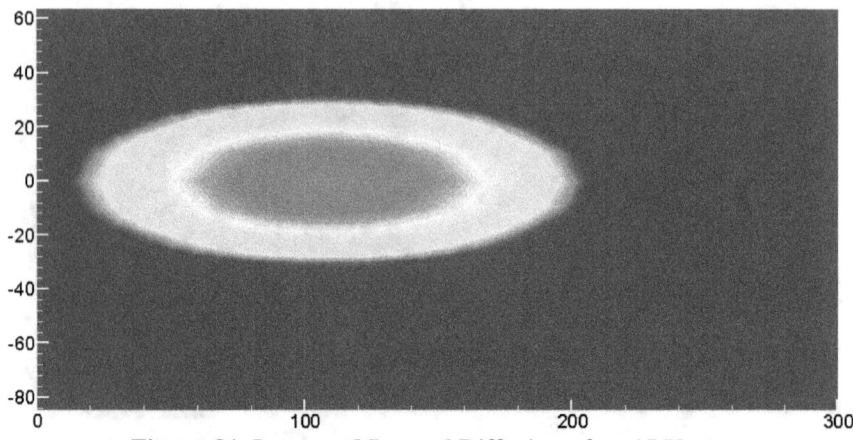

Figure 81. Increased Lateral Diffusion after 15 Years

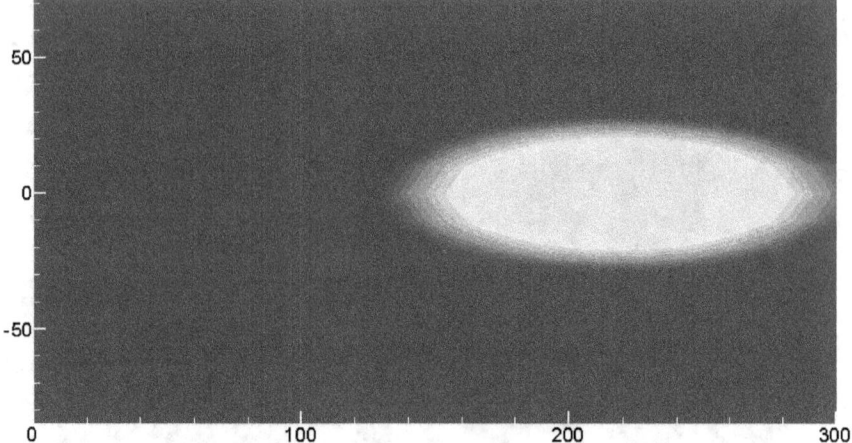

Figure 82. Increased Lateral Diffusion after 30 Years

If we add dispersion, we get:

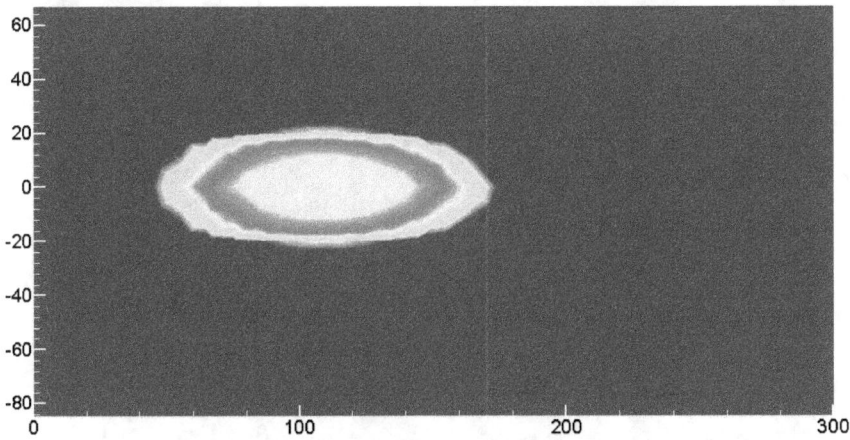

Figure 83. With Dispersion after 15 Years

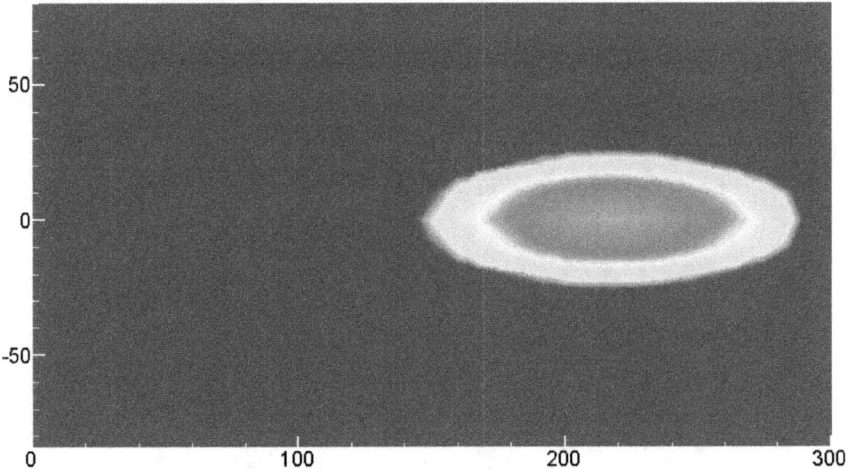

Figure 84. With Dispersion after 30 Years

Appendix A. Displaying Data in 3⁺D

Most any model spanning three spatial dimensions plus time—actually 4D—is capable of generating an enormous amount of data, requiring specialized software to display. In previous texts, I have discussed two such programs: TP2 and Tecplot™. I developed TP2 beginning in 1980 as TPLOT. TP2 (the second generation of TPLOT) is available free on the Web and can handle many types of data. Slices of a 3D field are shown in the figure below. The central view is from the top, the right view is looking in the side from the X direction, and the lower view is looking in the side toward the Y direction. The black crosshairs indicate the respective cutting planes, which are also listed in the lower right corner. TP2 can handle such data as text and also in binary, which decreases file size. We will generate both types of files for the 3D examples (see Appendices B and C for details).

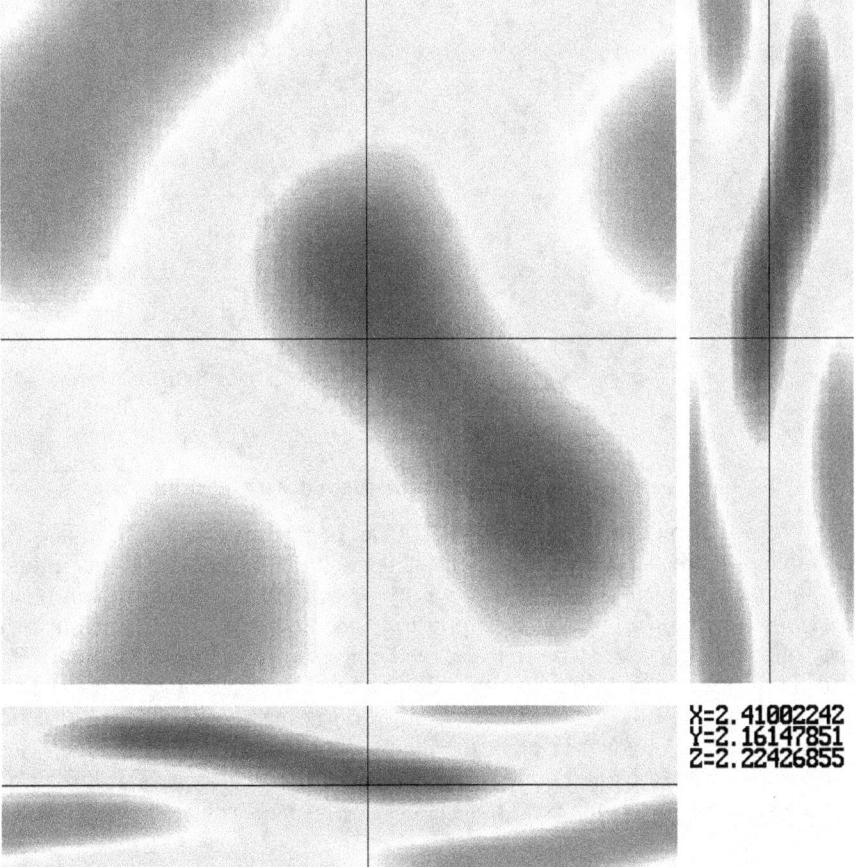

Figure 85. Slices of 3D Results in TP2

Tecplot™ is an excellent commercial tool[9] and can also handle many types of data. Besides the issues of cost (TP2 is free) and user support (TP2 has none), perhaps the biggest difference between TP2 and Tecplot™ is that the file extension (e.g., the ".dat" in myfile.dat) tells TP2 what type of data the file contains while headers within the file (whose extension is immaterial) provide this information to Tecplot™. This same data set viewed in Tecplot™ is shown below:

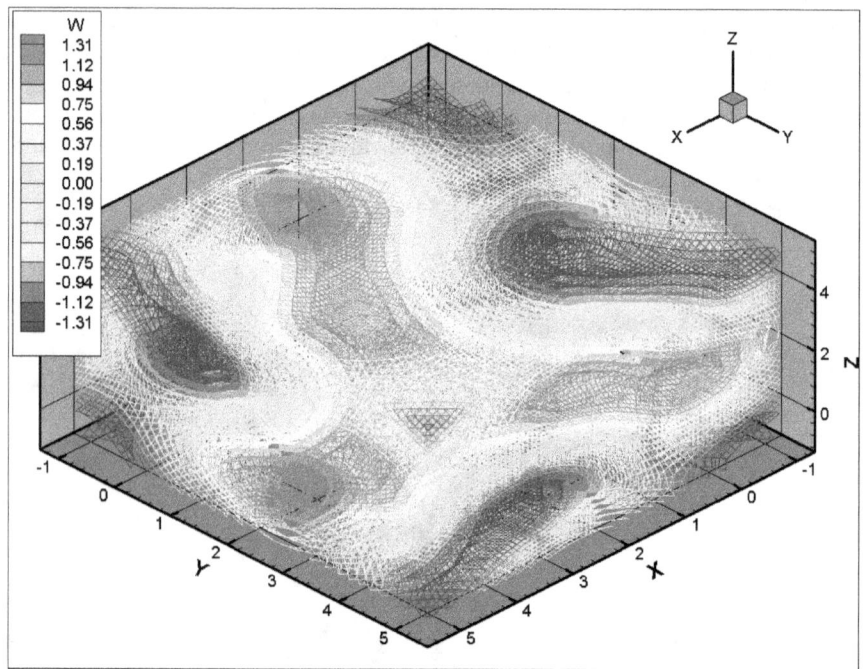

Figure 86. Same 3D Results Displayed with Tecplot™

We will also generate output files that can be read by Tecplot™. In our 2D examples we created the visualizations as the solution progressed through time but this isn't practical in 3D. In addition to the colorful representations in these previous two figures, it is often helpful to have additional information such as the boundary outline, air base runways, and cities shown in Figure 21. Both TP2 and Tecplot™ have the facility to incorporate such details in the respective layout files. This information in the form of a transparent overlay can also be embedded as an optional layer in the binary TP2 file (type MAP), which can also be compressed using Lempel-Ziv arithmetic encoding (type NM8) to make it even smaller. An example of a transparent overlay is shown in this next figure:

---

[9] This excellent product can be found at their web site https://www.tecplot.com/

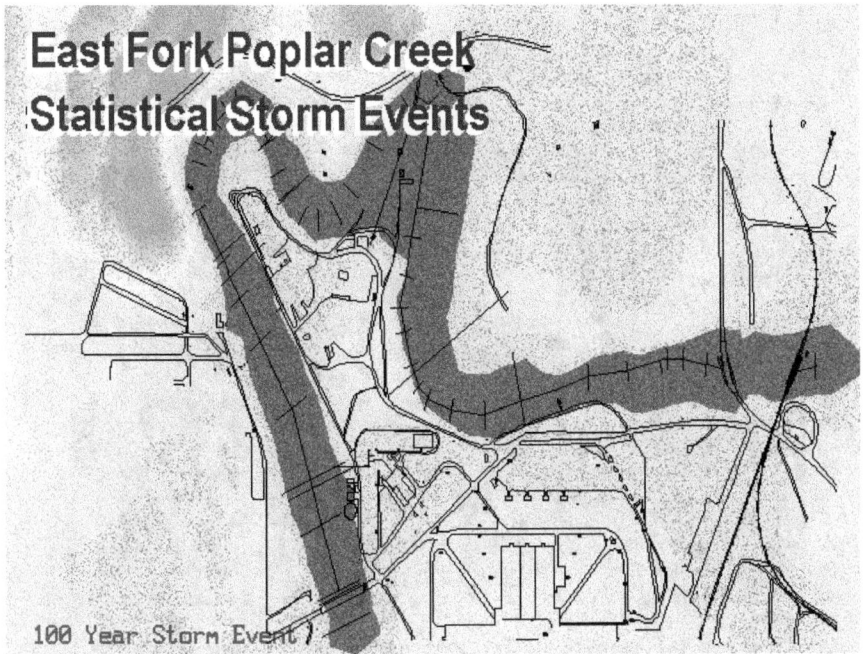

Figure 87. Example of Transparent Overlay Using TP2

Tecplot™ also provides transparent overlays. These are accomplished using what are called "geometries" within the layout file (name.LAY), an excerpt of which is listed below:

```
$!ATTACHGEOM
   ATTACHTOZONE = YES
   COLOR = RED
   FILLCOLOR = BLACK
   GEOMTYPE = LINESEGS3D
   PATTERNLENGTH = 4
   ARROWHEADANGLE = 20
   SCOPE = GLOBAL
   RAWDATA
1
14
870277 237001.25 100
870274.75 237014.25 100
870280.25 237152.75 100
870275.5 237170 100
870268.5 237178.75 100
870259 237184.5 100
870244.5 237186.75 100
870225.25 237186 100
870199.25 237041.5 100
```

A Tecplot™ transparent overlay is illustrated in this next figure:

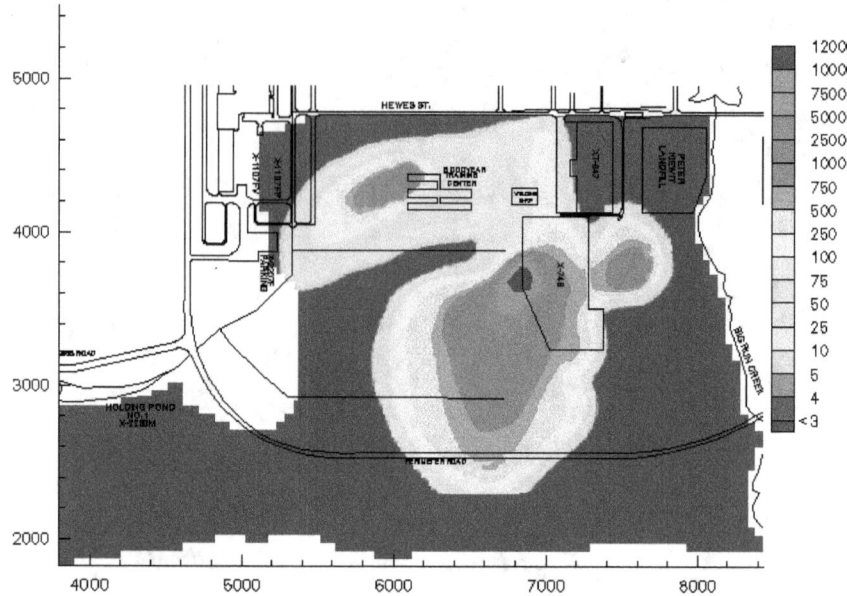

Figure 88. Example of Transparent Overlay Using Tecplot™

Appendix B. 3D Data files for Tecplot™

We will produce file directly for Tecplot™ but it may be useful to know that TP2 will convert a 3D file from its format to that of Tecplot™. Simply launch TP2, go to the menu, select "Convert", and then "volume => raw data". The resulting file can be loaded directly into Tecplot™ by using the menu to select "File" and then "Load DataFile(s)". The resulting file looks like this:

```
# converted 3D tabular data
# Nx=201, Ny=201, Nz=21
# -2500≤X≤2500
# -2500≤Y≤2500
# -25≤Z≤25
# -5.80914≤W≤4.82807
#  X    Y    Z    W
# this file can be read "as-is" by Tecplot
VARIABLES="X", "Y", "Z", "W"
ZONE I=201, J=201, K=21
-2500 -2500 -25 -5.80914
-2475 -2500 -25 -5.80914
-2450 -2500 -25 -5.80914
```

There is one "zone" or group of data, which has three dimensions, indicated by I, J, and K, having size 201x201x21. The headers are followed by 848,412 lines of X Y Z W. This particular file is 20,511,296 bytes in length. The original file in TP2 format was only 7,596,336 bytes in length, almost a factor of 3. For this particular format (IJK) the X, Y, and Z values are repetitious, whether evenly spaced or not. Filing the same values of X, Y, and Z over and over again is wasteful. Note that the file extension is immaterial.

This first format consists of regularly-spaced point (i.e., nodes). Tecplot™ will also accept several types of finite element data, including hexahedra (bricks). TP2 will also convert files to that format with menu convert volume => elements. This format looks like:

```
VARIABLES="X", "Y", "Z", "C"
ZONE N=99372 E=91575 F=FEPOINT ET=BRICK
109369 12051 359.8 0
109397 12071 360.6 0
109350 12079 360.3 0
etc.
110862 14745 701   0
110815 14753 701.2 0
110890 14764 700.9 0
 1 2  6 3 3823 3824 3828 3825
 2 5 10 6 3824 3827 3832 3828
 3 6 12 7 3825 3828 3834 3829
etc.
```

Appendix C: 3D Data Files for TP2

TP2 reads 3D regularly spaced data from a file having the extension TB3 (i.e., a 3D table of values). The format is quite simple: the number of X values, followed by the Xs, the number of Y values, followed by the Ys, the number of Z values, followed by the Zs, and the number of W values, followed by the Ws. For example:

```
201
-2500
-2475
-2450
...
201
-2500
-2475
-2450
...
21
-25
-22.5
-20
...
848421
-1.05735
0.0649466
2.53878
...
```

Appendix D. Build3D Model Builder

Build3D was developed to gather diverse information and build the corresponding input files to facilitate complex three-dimensional modeling, particularly for groundwater problems but also for other types of applications. Build3D reads in a variety of data, including topography, geological formations, soil types and depth, rivers and streams, water table, and recharge (rainfall). Each of these data sets must be in a different file and have a specific three-letter file extension (i.e., filename.ext).

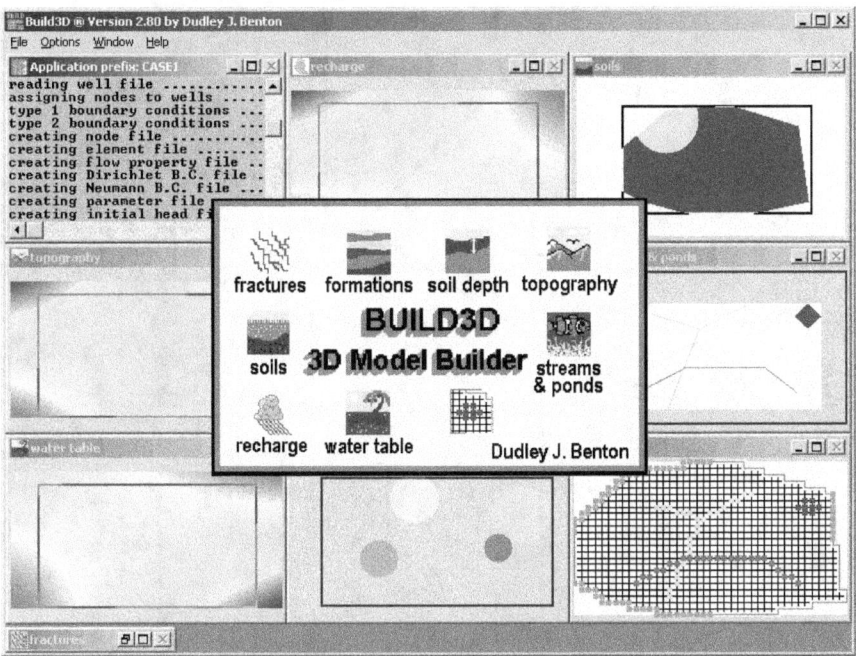

Figure 89. Typical Build3D Display

Build3D can create regular grids or optionally read in a prepared finite element grid. Build3D can handle finite elements (FEM) and finite differences (FDM). Three types of elements are recognized: hexahedra (bricks), pentahedra (prisms), and tetrahedra (three-sided pyramids).

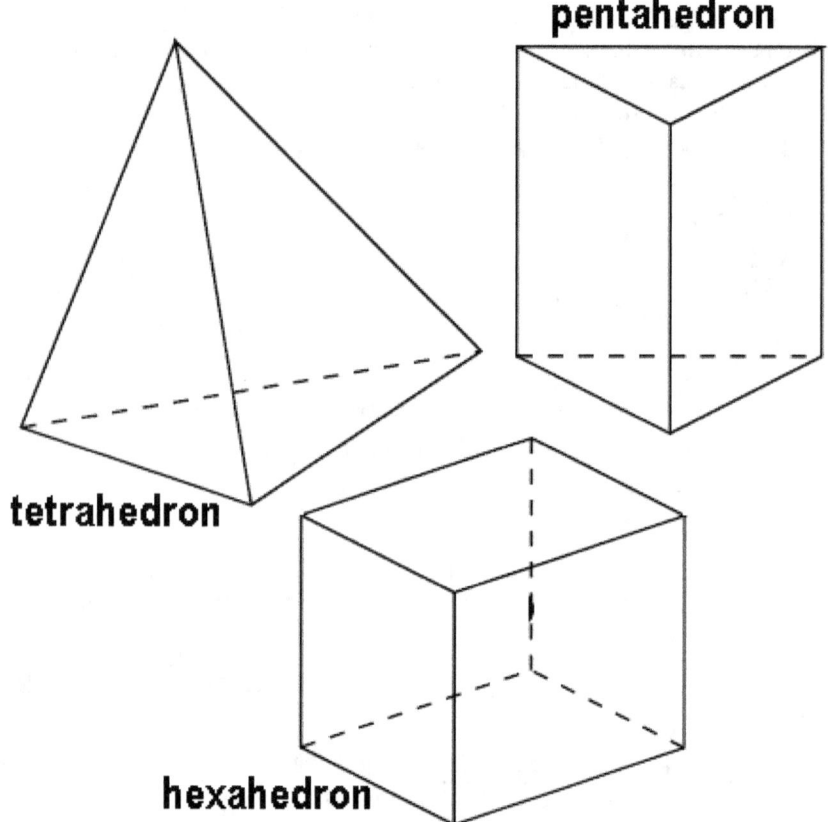

Figure 90. Three Types of Elements Recognized by Build3D

Hexahedral elements are shown in the lower right corner of the first figure in this appendix. Pentahedral elements are shown in the following figure:

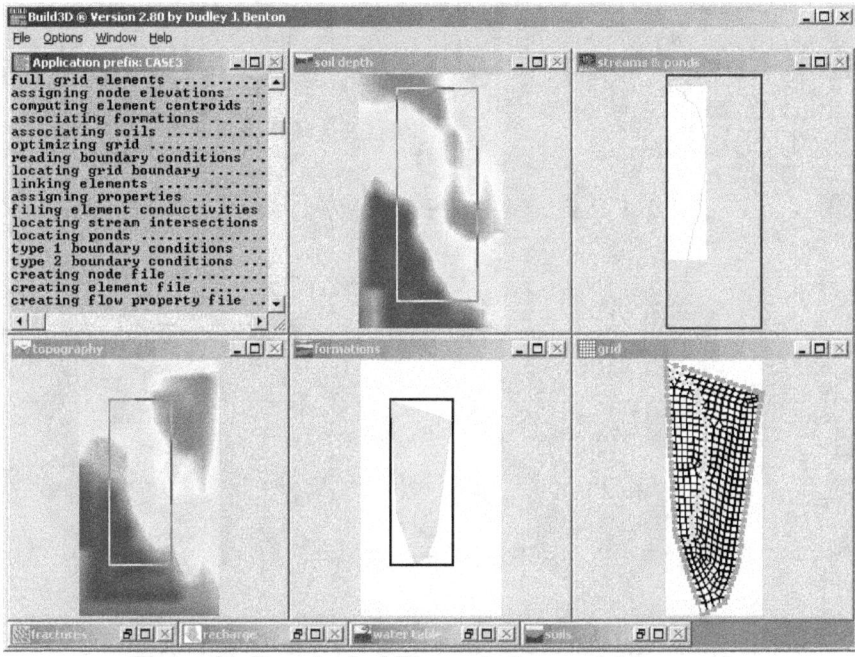

Figure 91. Pentahedral Elements (lower right corner)

Tetrahedral elements are shown in the "grid" box of this next figure:

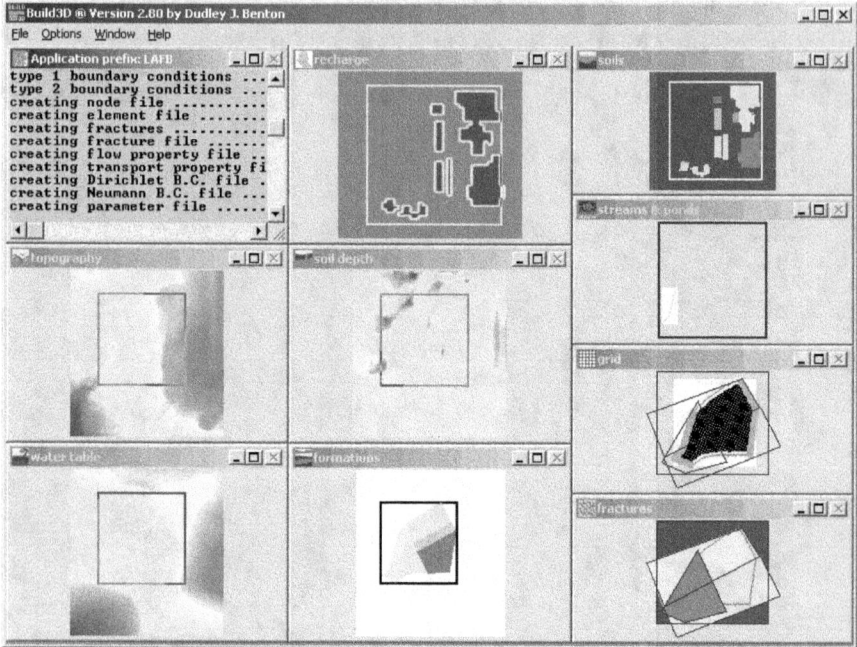

Build3D will create input files for MODFLOW and also FRAC3D. Build3D can be downloaded and used free of charge at the address below. Along with the executable and help file, three complete examples are included in the archive. Build3D also prepares input files for the particle tracker, PTRAX.

https://dudleybenton.altervista.org/projects/Build3D/index.html

Appendix E. Initial Concentrations

In the preceding examples we have used rather simple initial concentrations. In real like remediation projects, the initial concentrations can be quite complex. Most often data are only available at discrete locations, as sampling requires drilling wells and extraction. Special equipment is used to sample wells at controlled depth intervals to obtain vertical measurements. Once this data has been obtained, there are several ways to process it. Most often we need a 3D field of initial concentrations. Both TP2 and Tecplot™ have featured to handle this, including inverse distance interpolation and kriging.

One such example is shown in plan view below. The scattered points have been collapsed in depth and colored based on the log(concentration). The magenta polygon was the original estimated perimeter and the gray polygon is the expanded one. Changes like this were made throughout the project as more data was collected. This particular data were for TCE.

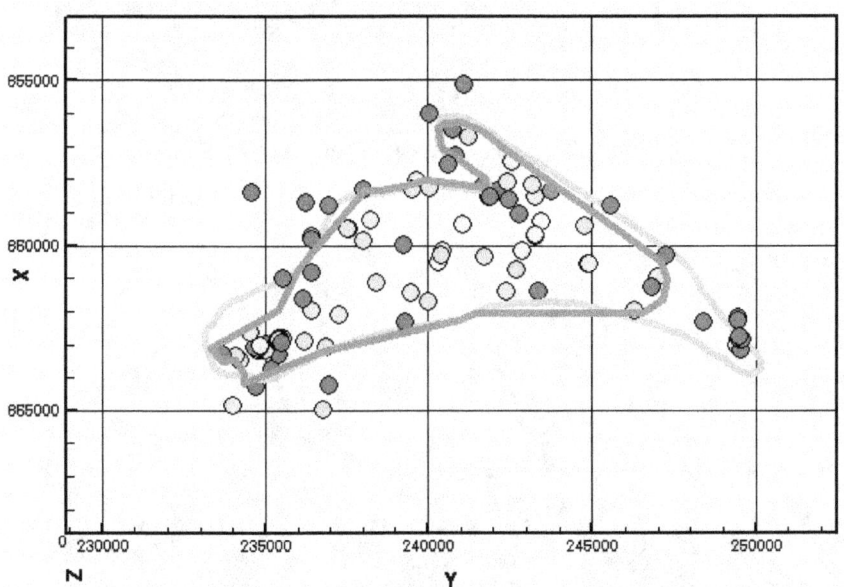

Figure 92. TCE Concentration Data from Multiple wells

This same data in 3D are shown in the next figure:

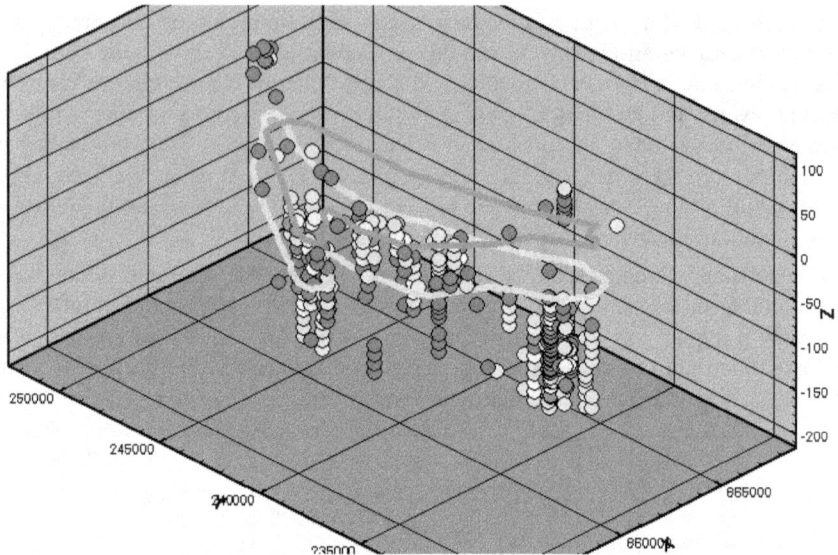

Figure 93. TCE Concentration Data in 3D

This same data shown with an interpolated plane slicing through the domain.

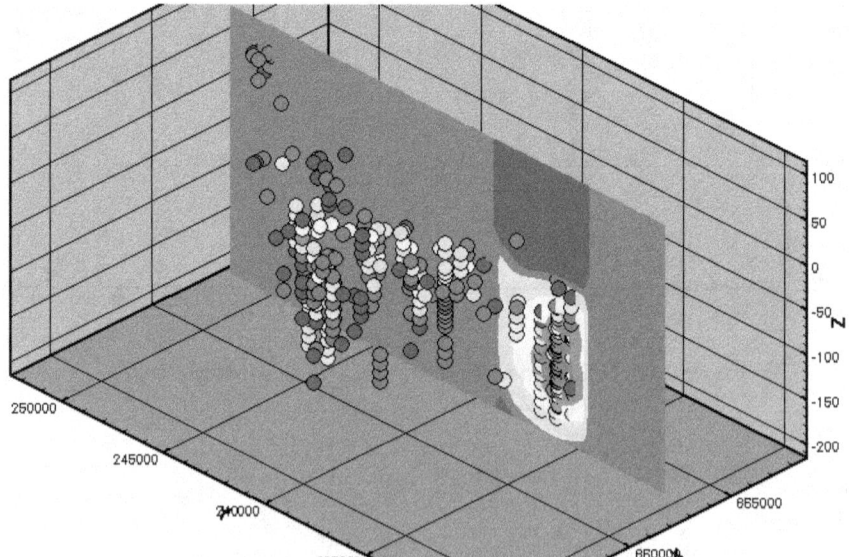

Figure 94. TCE Data with Slicing Plane

Data for a different substance (PCE) at the same location is shown in this next figure along with some initial contours.

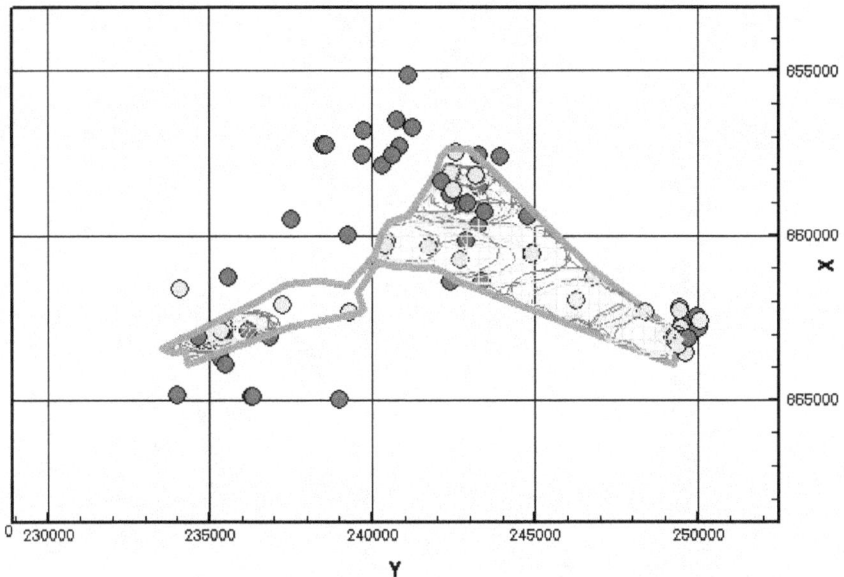

Figure 95. PCE Concentrations at the Same Wells

Both TP2 and Tecplot™ can interpolate and paint slices through the data in any of the three principle directions, which can be quite useful in evaluating and quantifying the initial conditions.

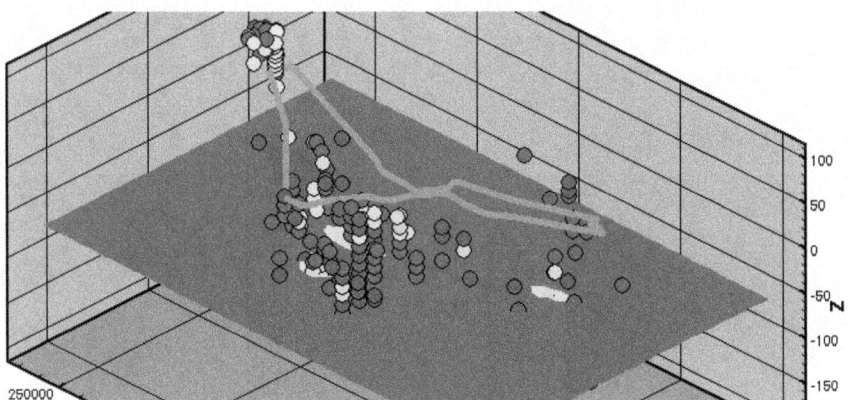

Figure 96. PCE Concentrations in 3D

also by D. James Benton

3D Articulation: Using OpenGL, ISBN-9798596362480, Amazon, 2021 (book 3 in the 3D series).

3D Models in Motion Using OpenGL, ISBN-9798652987701, Amazon, 2020 (book 2 in the 3D series).

3D Rendering in Windows: How to display three-dimensional objects in Windows with and without OpenGL, ISBN-9781520339610, Amazon, 2016 (book 1 in the 3D series).

A Synergy of Short Stories: The whole may be greater than the sum of the parts, ISBN-9781520340319, Amazon, 2016.

Azeotropes: Behavior and Application, ISBN-9798609748997, Amazon, 2020.

bat-Elohim: Book 3 in the Little Star Trilogy, ISBN-9781686148682, Amazon, 2019.

Boilers: Performance and Testing, ISBN: 9798789062517, Amazon 2021.

Combined 3D Rendering Series: 3D Rendering in Windows®, 3D Models in Motion, and 3D Articulation, ISBN-9798484417032, Amazon, 2021.

Complex Variables: Practical Applications, ISBN-9781794250437, Amazon, 2019.

Compression & Encryption: Algorithms & Software, ISBN-9781081008826, Amazon, 2019.

Computational Fluid Dynamics: an Overview of Methods, ISBN-9781672393775, Amazon, 2019.

Computer Simulation of Power Systems: Programming Strategies and Practical Examples, ISBN-9781696218184, Amazon, 2019.

CPUnleashed! Tapping Processor Speed, ISBN-9798421420361, Amazon, 2022.

Curve-Fitting: The Science and Art of Approximation, ISBN-9781520339542, Amazon, 2016.

Death by Tie: It was the best of ties. It was the worst of ties. It's what got him killed., ISBN-9798398745931, Amazon, 2023.

Differential Equations: Numerical Methods for Solving, ISBN-9781983004162, Amazon, 2018.

Equations of State: A Graphical Comparison, ISBN-9798843139520, Amazon, 2022.

Evaporative Cooling: The Science of Beating the Heat, ISBN-9781520913346, Amazon, 2017.

Forecasting: Extrapolation and Projection, ISBN-9798394019494, Amazon 2023.

Heat Engines: Thermodynamics, Cycles, & Performance Curves, ISBN-9798486886836, Amazon, 2021.

Heat Exchangers: Performance Prediction & Evaluation, ISBN-9781973589327, Amazon, 2017.

Heat Recovery Steam Generators: Thermal Design and Testing, ISBN-9781691029365, Amazon, 2019.

Heat Transfer: Heat Exchangers, Heat Recovery Steam Generators, & Cooling Towers, ISBN-9798487417831, Amazon, 2021.
Heat Transfer Examples: Practical Problems Solved, ISBN-9798390610763, Amazon, 2023.
The Kick-Start Murders: Visualize revenge, ISBN-9798759083375, Amazon, 2021.
Jamie2: Innocence is easily lost and cannot be restored, ISBN-9781520339375, Amazon, 2016-18.
Kyle Cooper Mysteries: Kick Start, Monte Carlo, and Waterfront Murders, ISBN-9798829365943, Amazon, 2022.
The Last Seraph: Sequel to Little Star, ISBN-9781726802253, Amazon, 2018.
Little Star: God doesn't do things the way we expect Him to. He's better than that! ISBN-9781520338903, Amazon, 2015-17.
Living Math: Seeing mathematics in every day life (and appreciating it more too), ISBN-9781520336992, Amazon, 2016.
Lost Cause: If only history could be changed..., ISBN-9781521173770, Amazon, 2017.
Mass Transfer: Diffusion & Convection, ISBN-9798702403106, Amazon, 2021.
Mill Town Destiny: The Hand of Providence brought them together to rescue the mill, the town, and each other, ISBN-9781520864679, Amazon, 2017.
Monte Carlo Murders: Who Killed Who and Why, ISBN-9798829341848, Amazon, 2022.
Monte Carlo Simulation: The Art of Random Process Characterization, ISBN-9781980577874, Amazon, 2018.
Nonlinear Equations: Numerical Methods for Solving, ISBN-9781717767318, Amazon, 2018.
Numerical Calculus: Differentiation and Integration, ISBN-9781980680901, Amazon, 2018.
Numerical Methods: Nonlinear Equations, Numerical Calculus, & Differential Equations, ISBN-9798486246845, Amazon, 2021.
Orthogonal Functions: The Many Uses of, ISBN-9781719876162, Amazon, 2018.
Overwhelming Evidence: A Pilgrimage, ISBN-9798515642211, Amazon, 2021.
Particle Tracking: Computational Strategies and Diverse Examples, ISBN-9781692512651, Amazon, 2019.
Plumes: Delineation & Transport, ISBN-9781702292771, Amazon, 2019.
Power Plant Performance Curves: for Testing and Dispatch, ISBN-9798640192698, Amazon, 2020.
Practical Linear Algebra: Principles & Software, ISBN-9798860910584, Amazon, 2023.
Props, Fans, & Pumps: Design & Performance, ISBN-9798645391195, Amazon, 2020.
Remediation: Contaminant Transport, Particle Tracking, & Plumes, ISBN-9798485651190, Amazon, 2021.

ROFL: Rolling on the Floor Laughing, ISBN-9781973300007, Amazon, 2017.
Seminole Rain: You don't choose destiny. It chooses you, ISBN-9798668502196, Amazon, 2020.
Septillionth: 1 in 10^{24}, ISBN-9798410762472, Amazon, 2022.
Software Development: Targeted Applications, ISBN-9798850653989, Amazon, 2023.
Software Recipes: Proven Tools, ISBN-9798815229556, Amazon, 2022.
Steam 2020: to 150 GPa and 6000 K, ISBN-9798634643830, Amazon, 2020.
Thermochemical Reactions: Numerical Solutions, ISBN-9781073417872, Amazon, 2019.
Thermodynamic and Transport Properties of Fluids, ISBN-9781092120845, Amazon, 2019.
Thermodynamic Cycles: Effective Modeling Strategies for Software Development, ISBN-9781070934372, Amazon, 2019.
Thermodynamics - Theory & Practice: The science of energy and power, ISBN-9781520339795, Amazon, 2016.
Version-Independent Programming: Code Development Guidelines for the Windows® Operating System, ISBN-9781520339146, Amazon, 2016.
The Waterfront Murders: As you sow, so shall you reap, ISBN-9798611314500, Amazon, 2020.
Weather Data: Where To Get It and How To Process It, ISBN-9798868037894, Amazon, 2023.

www.ingramcontent.com/pod-product-compliance
Lightning Source LLC
Chambersburg PA
CBHW071521220526
45472CB00003B/1112